How Birds Evolve

How Birds Evolve

WHAT SCIENCE REVEALS ABOUT THEIR
ORIGIN, LIVES, AND DIVERSITY

DOUGLAS J. FUTUYMA

PRINCETON UNIVERSITY PRESS
PRINCETON & OXFORD

Published by Princeton University Press
41 William Street, Princeton, New Jersey 08540
99 Banbury Road, Oxford OX2 6JX

press.princeton.edu

All Rights Reserved

First paperback printing, 2024
Paperback ISBN 9780691264639

The Library of Congress has cataloged the cloth edition as follows:

Names: Futuyma, Douglas J., 1942– author.
Title: How birds evolve : what science reveals about their origin, lives, and diversity / Douglas J. Futuyma.
Description: Princeton : Princeton University Press, [2021] | Includes bibliographical references and index.
Identifiers: LCCN 2021008373 (print) | LCCN 2021008374 (ebook) | ISBN 9780691182629 (hardback) | ISBN 9780691227269 (ebook)
Subjects: LCSH: Birds—Evolution. | BISAC: SCIENCE / Life Sciences / Zoology / Ornithology | SCIENCE / Life Sciences / Biology
Classification: LCC QL677.3 .F88 2021 (print) | LCC QL677.3 (ebook) | DDC 598–dc23
LC record available at https://lccn.loc.gov/2021008373
LC ebook record available at https://lccn.loc.gov/2021008374

British Library Cataloging-in-Publication Data is available

Editorial: Alison Kalett and Whitney Rauenhorst
Production Editorial: Kathleen Cioffi
Jacket/Cover Design: Lauren Smith
Production: Jacqueline Poirier
Publicity: Matthew Taylor and Kate Farquhar-Thomson
Copyeditor: Eva Silverfine

Jacket/Cover images (clockwise from top left): Elliot's Bird-of-paradise (*Epimachus ellioti*), Natural History Museum, London / Alamy; Amethyst Woodstar (*Calliphlox amethystina*), Florilegius / Alamy; "Toucan No. 2," illustration from *Histoire naturelle des oiseaux de paradis et des rolliers*, by François Levaillant, engraved by Jacques Louis Pérée, 1801–6 / The Stapleton Collection / Bridgeman; Great Crested Flycatcher (*Myiarchus crinitus*), Encyclopaedia Britannica / UIG / Bridgeman; Eurasian Nuthatch (*Sitta europaea*), Old Images / Alamy

This book has been composed in Arno

Color versions of the images following p.36 can be accessed at:
https://press.princeton.edu/books/paperback/9780691264639/how-birds-evolve

CONTENTS

PREFACE

This book is for birders and for all who enjoy nature and sometimes ask questions about what they see. Why are male birds more colorful in some species but not others? Why do some owls and other species come in different colors, independent of sex or age? Why is parental care of young the duty of females in some species, males in others, and both parents in still others? What are species, how do they arise, and why are there so many more in tropical regions than in the temperate zone? How and when did the astonishing variety of birds evolve? These questions may also appeal to people who are interested in evolution, if not in birds, and I hope they, too, will find something of value in this book.

All these topics, and many more, are the province of evolutionary biology because all species and all their characteristics are products of a history of evolutionary origin and modification. Many other books have focused on specific aspects of bird evolution, such as their origin from dinosaurs, but my aim is to describe the light that evolutionary science casts on diverse aspects of birds' lives and diversity.

As for my own journey, I became fascinated by animals as a boy and started birding at about age eleven in the parks of New York City. My interest in birds and other animals led to majoring in biology at Cornell University, where I focused on evolution and ecology. My doctoral research at the University of Michigan combined these areas in an experimental study of competition between fruit flies. Since joining the faculty at Stony Brook University (in Long Island, New York) in 1970, I developed and have taught undergraduate and doctoral courses in evolutionary biology and, occasionally, ecology and entomology. I did research for many years on the evolution of interactions between plant-feeding insects and their food plants and have also published scientific papers on speciation, the evolution of ecological specialization, constraints on evolution, and other topics. I wrote two successful textbooks of evolutionary biology that together have gone through seven editions (so far). I hasten to say, though, that I have done no research on birds and have no professional qualifications in ornithology. I can only read about research on bird evolution, and I am an enthusiastic birder, having watched birds in fifty countries (so far). Writing this book has been an education and a great pleasure, as it joins my professional and avocational interests.

You will meet a great many bird species in this book because so many have been subjects of important research, because diversity itself is the focus and calls for explanation, and because readers in different regions or countries will be familiar with different species. Unfortunately, it is possible to illustrate only a few species. So I encourage readers to find images of unfamiliar species online; a good source (although by subscription) is birdsoftheworld.org, by the Cornell Lab of Ornithology. The print equivalent is *All the Birds of the World* (del Hoyo 2020). For bird names, I generally follow the IOC (International Ornithologists' Union) World Bird List (www.worldbirdnames.org). I provide the scientific name of a species when it is first mentioned because some species go by different English names in different books and checklists.

Although some technical terms are necessary for brevity, I have tried to use as few as possible and to explain concepts in simple terms. Much of evolutionary biology (and of science generally) involves making predictions about what we ought to see if a hypothesis is right and comparing data with the predictions. I hope the ideas and how they do or don't match data about birds are straightforward, but in places they may need a little extra attention. I have written most of the book as if in conversation, trying to avoid professorial lecturing, and occasionally I recount my birding experiences. I see birds through the eyes of a biologist, but the emotional and esthetic experience of birding is foremost for me, and I think for most readers.

Acknowledgments

I am immensely grateful to Lucille Betti-Nash and Stephen Nash for their professional illustrations of birds in the text figures. I thank Ross Aftel and Dean Bobo for help with digital challenges, Patrice Domeischel for help with photos, Noah Strycker for checking bird names, and Sue and Ken Feustel, Kevin Padian, and Richard Prum for advice. Alice Deutsch, Mihai Chitulescu, Patricia Lindsay, Shaibal Mitra, Eric Ozawa, Phil Ribolow, Andrew Rubenfeld, and Roy Tsao kindly reviewed parts of the manuscript, and Andrew Moore provided particularly important reviews of chapter 3. For providing information, I am grateful to colleagues Resit Akçakaya, Joel Cracraft, Scott Edwards, Bob

Holt, Bette Loiselle, Támas Székely, Morgan Tingley, Frank Sulloway, and Hamish Spencer. I am grateful to Rob DeSalle for sponsoring me as a Research Associate at the American Museum of Natural History and for providing desk space, and I have enjoyed enlightening conversations with museum staff, including George Barrowclough, Paul Sweet, and Joel Cracraft, who also provided access to books in his library. Some of the color photos were generously provided by friends, birding guides, and professional acquaintances, including Ciro Albano, John Barkla, Dǔsan Brinkhuizen, Nick Davies, David Erterius, Doug Gochfeld, Rich Hoyer, Hannu Jännes, Phil Jeffrey, Mark Kirkpatrick, Markus Lilje, Daniel López Velasco, Bruce Lyon, Lisa Nasta, Glenn-Peter Saetre, Bryan Shirley, Thomas B. Smith, Michael Stubblefield, and Steve Walter. Andrew Cockburn, Mike Cooper, Simon Griffith, Alan Krakauer, and Chris Lester helped find sources of some photos; Matt Medler and Michael Webster, at the Macaulay Library of the Cornell Lab of Ornithology, arranged for some photos from eBird contributors; and other photos were kindly provided by Bjorn Aksel Bjerke, Michael Fidler, Tobias Hayashi, Jon Irvine, Patrik Karell, Miroslav Kral, James Mott, and Shailesh Pinto. I am grateful to Doug Gill for rekindling my interest in active birding, to many friends and companions in the birding community, and to the students and professional colleagues who have enriched my life. I apologize to those whom I may have inadvertently failed to acknowledge.

I appreciate the helpful suggestions of three anonymous reviewers, and I am grateful for the advice, support, and forbearance of personnel at Princeton University Press, especially Alison Kalett, Abigail Johnson, Whitney Rauenhorst, Lisa Black, and Kathleen Cioffi.

Douglas J. Futuyma
Stony Brook, New York, November 19, 2020

How Birds Evolve

1

In the Light of Evolution

BIRDS AND EVOLUTIONARY SCIENCE

A few years ago, I joined a birding tour of Ghana. After several days of enjoying such exotic species as drongos, hornbills, and pratincoles, we encountered a beautiful red and black finch, the Black-bellied Seedcracker (*Pyrenestes ostrinus*).[1] I was delighted to see this species because I had long known, and had described in my textbook of evolutionary biology, a study of this species by Thomas Smith,[2] a professor at University of California–Los Angeles. Smith had followed the life of members of a population in Cameroon by fitting each individual with a unique combination of colored leg bands. Bill size is highly variable in seedcracker populations; most birds have either small or large bills, although a minority are intermediate (plate 1). Smith found that large-billed birds feed more efficiently on the large, hard seeds of one species of sedge and small-billed birds handle the small seeds of another sedge more efficiently. Large-billed and small-billed birds both had higher rates of survival than intermediate birds: a striking example of natural selection in action. By occupying somewhat different "ecological niches," birds with different genotypes (specific combinations of genes) persist. Years later, when the study of genomes had advanced, Smith and his collaborators determined that the inherited difference in bill size is caused by different forms of a single gene (called *IGF1*, or insulin-like growth factor 1).[3] As we admired the seedcracker, I told my companions this story. One of them exclaimed, "So that's why it doesn't look like the picture in the field guide! I wondered if the book was wrong." He was intrigued by the idea that different members of a species have different diets and ways of life.

Some birders are focused on seeing and listing species; others are curious about the lives and features of the birds they see. Once in a while a fellow birder, knowing that I'm a biologist, will ask me a question. Sometimes it is along the lines of "how can birds fly so fast through dense vegetation without hitting it?" or "how can a tiny Blackpoll Warbler (*Setophaga striata*) fly non-stop from New England to Venezuela?" I awkwardly answer that I don't know much about how birds achieve these amazing feats because those are topics studied by biologists who specialize in bird physiology or brain function, and I haven't followed those fields since I was a student. Some other questions, though, tempt me to say more than they may want to hear. (And I can resist anything but temptation.[4]) Why do some bird species have different color morphs? Why are males more brightly colored than females in some species but not others? Why do albatrosses and many other sea birds lay only one egg? How come I can see more bird species in a two-week birding tour in Peru than in an entire year in eastern North America? Why do they keep changing bird classifications, and how do they know falcons are closer to parrots than to hawks?

Most questions about birds fall into two categories—*how* and *why*—that correspond to two major kinds of biological research. Much of biology poses "how" questions: it aims to understand how organisms function—how the molecular, cellular, and organ components of an organism work, here and now, without reference to how they came to be. "Why" questions are the province of evolutionary biology. We ask why a Eurasian Golden Oriole (*Oriolus oriolus*) or an American Baltimore Oriole (*Icterus galbula*) is brightly colored because we understand that it could have been otherwise: something in its history—in its evolution—caused it to be bright rather than drab. For every characteristic of every species, we can ask "how" questions about its functional role (if any) in an organism's lifetime, complemented by "why" questions about its origin. All species of birds have evolved from a single ancestral species ("common ancestor"), which was one of a great many species of vertebrates that all evolved from a single, more ancient, common ancestor; this, in turn, was a descendant of the ur-ancestor of all animals, from sponges to primates. And so every feature of every bird, from its DNA sequences to its behaviors, has come into existence—has evolved—during this history of descent.

Evolutionary biologists attempt to develop broad principles that can explain all these features of all species. Evolutionary biology illuminates every area of biological research and every group of organisms. The geneticist Theodosius Dobzhansky, who helped to shape modern evolutionary biology, rightly wrote that "nothing in biology makes sense except in the light of evolution."[5] There are biologists who study biochemical processes within cells and biologists who study how these processes evolved—and likewise for the structure and function of genomes, brains, and hormones. Among ornithologists, some take a mostly functional approach, and others a more evolutionary approach, to bird physiology, morphology,[6] behavior, and life histories. Others are devoted to understanding the history of bird evolution—how and when birds' form, behavior, habitat use, and geographical distribution diversified during their descent from their common ancestor. The amount of research that bears on bird evolution is immense: when I entered "evolution and bird*" in a search engine (Web of Science), it yielded 73,200 articles in scientific journals.[7] Variant search terms would add many more.

So for almost any question we might ask about how birds evolved, there is plenty of research on which to draw. Nevertheless, the known is far less than the unknown. Questions such as "how do new species form?" and "why do female birds prefer flashy males?" are debated and are the subjects of active research. And while we may be able to provide a general answer to a question (e.g., why do birds' bills differ in shape?), there may not be a definitive answer for a particular species. (I don't know of any research about why the bill of the Groove-billed Ani [*Crotophaga sulcirostris*] is grooved.) Evolutionary biologists strive, instead, to develop theories that should apply to a wide range of species but which require detailed information to explain particular cases. For example, there are several models[8] to account for genetic polymorphism—the persistence of two or more genetically different types within a population, such as the color "phases" of the Tawny Owl (*Strix aluco*) and the Eastern Screech Owl (*Megascops asio*). Information about the survival and reproduction of each form, under several environmental conditions, may be needed to match a particular instance to one of the models.

I can imagine someone thinking, at this point, "I watch birds because I'm entranced by their beauty and their behavior or because I enjoy the challenge of finding and identifying as many species as I can. It's an aesthetic, emotionally rewarding experience. Doesn't looking at a bird with the cold analytical eye of science ruin the experience?" Of course, I can't speak for everyone, but for me, birding certainly has those rewards, and the more I know, the more my appreciation is enhanced. As many as I have seen, I still am overwhelmed by a peacock's beauty, but it also spurs me to ask why and how it came to be, and having an answer enlarges and makes whole my experience. We integrate intellectual and aesthetic appreciation when we want to know the names of the birds we encounter and to which family or group a species belongs.

With knowledge of their biology, the most common, everyday birds take on new interest. Take the ubiquitous House Sparrow (*Passer domesticus*).[9] When I stop to look at a House Sparrow, I sometimes think of its broader evolutionary context: other species in the genus *Passer*. For example, the Italian Sparrow (*Passer italiae*) originated as a hybrid between House and Spanish Sparrows (*Passer hispaniolensis*) (see chapter 10), and the Eurasian Tree Sparrow (*Passer montanus*) replaces the House Sparrow as a human associate in southeastern Asia. The House Sparrow itself shows interesting geographical variation in Europe: northern birds are bigger than birds in the south. This is one of many species of birds and mammals that have this pattern due to adaptive evolution: larger bodies lose heat more slowly than smaller ones and are advantageous in colder regions. What is more, since House Sparrows were introduced from Europe into North America in 1851, they have spread widely, and northern populations have evolved larger size. This was one of the first examples of how rapid evolution can be; Darwin never imagined that evolutionary changes could happen within a few human lifetimes.

The Superb Fairywren (*Malurus cyaneus*) in Australia (plate 2) is another example of a common bird that poses interesting questions. A group usually has two or more bright blue and black males and several brown birds that include both males and females. Biologist Andrew Cockburn and his associates studied the extraordinary breeding behavior of fairywrens for more than twenty-five years.[10] The bright-plumaged and brown males all cooperate to rear nestlings. Cooperative breeding is known in many birds, and why it has evolved poses a very interesting question (chapter 7). But there is more: female fairywrens, to a greater extent than any other bird yet known, engage in "extra-pair copulation," or adultery: they will travel across intervening territories to mate with a "hotshot" male. The female's male associates dutifully help raise babies that usually aren't their own offspring. Why are females so unfaithful, and why do males stay and rear the offspring?

These are fascinating questions that evolutionary biology can help to answer—as it can shed light on countless other aspects of birds, ranging from their coloration and structure to their geographic distribution and diversity. My aim in this book is to pose such questions and show how insights from evolutionary biology can answer them. Also, research into these topics has revealed features of many species that I think will amaze and delight anyone who likes birds and help them appreciate birds all the more. And if some readers learn more about evolution and how it is studied, the book will have served another purpose—sharing some of the richness of evolutionary science that I have found so rewarding.

———

By "evolution," biologists usually mean change in the features of a single species over time (that is, across generations) as well as the division of a single species into two or more descendant species, both of which undergo change. The alterations of a feature must be inherited to count as evolutionary change. Some features can be affected by an individual's environment, but these changes are generally not inherited. A generation of people might be lighter skinned than their grandparents because they work in offices instead of fields and so are less suntanned, but this doesn't count as evolution. As inheritance is a defining feature of evolution, evolutionary change of organisms' features (their *phenotype*) is accompanied by evolution at the level of the genes. There is also evolution at the genetic (DNA) level that may not affect any features of the organism.

In *The Origin of Species*, Darwin developed two main themes: that all living things have descended, with modification, from common ancestors; and that the chief cause of modification is natural selection of inherited variations. The wealth of insights, hypotheses, and information in Darwin's writings is staggering. Every time I read a few pages of *The Origin of Species*, I'm simply floored by the questions he thought to ask, the possible answers he advanced, and the evidence he found in an extraordinary range of facts, some of them seemingly trivial. During his voyage on the *Beagle*, he notices, in South America, that a flycatcher, the Great Kiskadee (*Pitangus sulphuratus*) (figure 1.1), sometimes acts like kestrels and kingfishers when foraging. Later he cites this, in *The Origin of Species*, to illustrate that species might change and perhaps become adapted to new ways of life. Not everyone can see a world in a grain of sand, but Darwin realized that a coherent explanation or theory must be able to accommodate, and build on, every fact, however trivial it might seem.

Evolutionary biology today is devoted to Darwin's two great themes: what has happened in the evolution of the world's organisms, and what have been the causes of these evolutionary events?

In studying the history of evolution, biologists today draw mostly on two sources of information (the subject of chapters 2 and 3). One is the fossil record. The other is the similarities and differences among living species in their characteristics and DNA sequences. This information enables biologists to

FIGURE 1.1. A Great Kiskadee (*Pitangus sulphuratus*), a common flycatcher in much of tropical America, north to the border of Texas. (Art, Luci Betti-Nash.)

piece together species' relationships, to infer their family tree, or phylogeny (chapter 2). Both phylogenies and fossils can yield information on how features have changed; for instance, they tell us that flightless birds like kiwis and penguins have evolved from flying ancestors and that the same transition has happened independently in kiwis, penguins, and many other lineages. Often, such phylogenetic information can help us understand how certain features that differ among species, such as bill shape, are adaptive.

How does evolution happen? Darwin's greatest idea, one of the most important ideas in human history, was natural selection. (The philosopher Daniel Dennett called evolution by natural selection "the single best idea anyone has ever had."[11]) *If* a character (meaning a feature or trait) varies among individuals of a species, and *if* the variation is at least partly hereditary (i.e., genetic), and *if* individuals with a certain variant condition tend to survive or

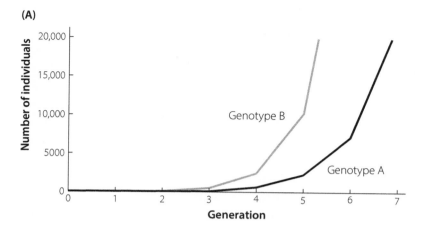

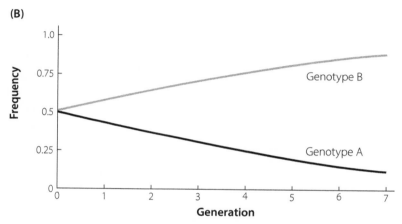

FIGURE 1.2. A simple model of genetic change by natural selection. Two genotypes (groups of organisms with specific combinations of genes) differ in a characteristic that affects survival or reproduction. Genotype A has an average fitness equal to 3, meaning than an average newborn A leaves 3 offspring. Genotype B has an average fitness of 4, because it is more likely to survive and reproduce or because females lay more eggs. In the upper diagram, the number of B individuals grows faster than the number of As. It therefore makes up an increasing proportion (frequency) of the population, as shown in the lower diagram. (From Futuyma and Kirkpatrick 2017.)

reproduce more than others, *then* the proportion of that variant type in the species population will increase from one generation to the next, and it may ultimately replace all other variants (figure 1.2). Natural selection, then, is simply an average difference in the survival and reproduction of genetically different types of organisms. Darwin postulated that this process is the chief cause of evolution, and certainly of adaptive evolution—the origin and alteration of characteristics that enhance survival and reproduction. He likened natural selection to human selection of domesticated animals and plants, in which breeders propagate their stock from individuals that have particularly desirable features.

The Origin of Species was first published in 1859. Seven years later, an obscure monk, Gregor Mendel, published an obscure paper on inheritance in peas that was not widely noticed until 1900, when it became the foundation of the modern science of genetics. Since then, genetic knowledge has become the chief framework for describing the processes of evolution within species. This framework, expressed in both words and equations, describes the factors that cause genetic changes in species. In the simplest terms: a new version of a gene (an *allele*) comes into existence by *mutation* (usually a change of one of the units of a DNA sequence). At first it is very rare—only one or a few individuals carry the allele. If this allele alters a characteristic in such a way as to increase an individual's chance of survival or reproduction, it is said to be *naturally selected* and may become more common because such individuals survive or reproduce more than those that lack the allele and the advantageous feature. Perhaps the allele entirely replaces the original form of the gene (the new allele is *fixed*), and the population as a whole has a somewhat altered *phenotype* (i.e., characteristic: shorter legs, differently colored bill, different display behavior—whatever feature the gene affects).

Two of the factors that affect genetic evolution are mutation and natural selection. But there are others. Suppose a local population of the species is flooded with immigrants from another population with a different allele, and the immigrants interbreed with the residents. The proportion (or *frequency*) of the residents' original allele is lower and the frequency of the immigrants' allele is higher than before. This process, called *gene flow*, can change a population's genetic composition. Finally, and very importantly, the frequencies of two alleles (say, old allele and new mutation) are affected by pure chance.[12] Some individuals suffer accidental deaths, or are unlucky in love, no matter how genetically vigorous and reproductively potent they are—and their failure to pass on their genes changes the allele frequencies in the next generation, however slightly. This purely random change is called *genetic drift*. Over the course of generations, the frequency of an allele will fluctuate, and since there is no reason for the ups to precisely equal the downs, the frequency will eventually go to 0.0 or 1.0: the allele will be lost altogether or it will completely replace other alleles (figure 1.3). If the population is very small, each individual's bad versus

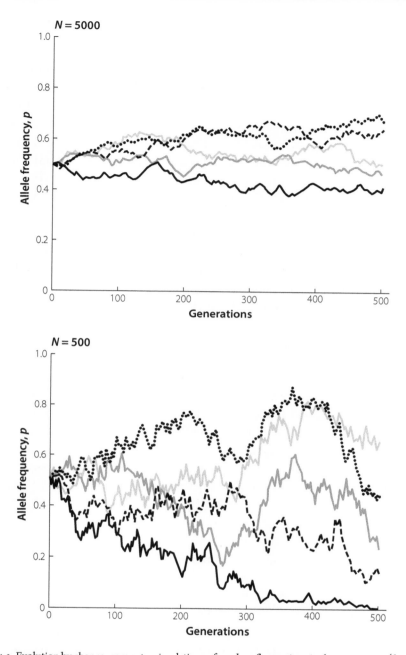

FIGURE 1.3. Evolution by chance: computer simulations of random fluctuations in the proportion (frequency) of one of two alleles (forms of a certain gene). Individuals that carry one allele or the other do not differ in fitness. In the top diagram, five populations, each of 5,000 individuals in each generation, evolve over the course of 500 generations. The frequency of one of the alleles is shown to fluctuate in each of the five populations, which come to differ even though all started with the same frequency (with both alleles equally common). The bottom diagram shows the same kind of history, but the populations are smaller (500 individuals in each generation). The fluctuations are greater, and the populations become different faster. In one of the populations, the one allele has dropped to zero frequency—so the other allele has reached a frequency of one (100% of the gene copies). That allele has taken over that population not because it is better but because it was lucky. (From Futuyma and Kirkpatrick 2017.)

good luck will have a bigger impact than if the population is large. So genetic drift changes allele frequencies faster in small than in large populations. Now suppose a particular allele enhances the chance of survival (it is advantageous) but only slightly. (Selection is said to be weak.) Both natural selection and genetic drift are operating, and if the population is small enough, random drift will be more influential than weak natural selection, and the advantageous allele may not become a fixed feature of the population. Whether natural selection or genetic drift rules depends on the strength of natural selection compared with the population size.

Mutation, natural selection, genetic drift, and gene flow affect evolution within the various local populations of a species and in a species as a whole. When gene flow between populations of a species is curtailed, the other three processes continue more independently in each of the separated populations, enabling them to become more different from each other. Under some conditions, the populations may ultimately become different species (chapter 10).

These processes underlie evolutionary changes within a species; they are very generic (and genetic) ideas that can be used to describe changes in everything from DNA sequences to biochemical, anatomical, and behavioral characteristics. And these concepts pervade all of evolutionary biology, including phylogenetic and paleontological studies of the history of evolution of birds (and everything else). A lot of evolutionary research involves trying to interpret differences within and among species in these terms. Chapters 4 and 5 include some fascinating examples of how biologists try to study these factors, especially natural selection, as they apply to specific characteristics of birds.

————

Birds have been and continue to be immensely important in the development of evolutionary science.[13] To be sure, they have been less useful than insects, plants, and bacteria in the study of the genetic foundations of evolution, partly because they don't reproduce as rapidly (although they are playing a larger role in genetic studies today with the growth of genomics). But birds have contributed more to studies of the evolution of physical characteristics, behavior, life histories, ecology, speciation, and geographic distribution than almost any other major group of organisms. My examples start with Darwin (of course!). In *The Voyage of the Beagle* (1839),[14] he refers to more than fifty species of birds that drew his attention, some of which he later used as evidence in *The Origin*

1. Geospiza magnirostris.
2. Geospiza fortis.
3. Geospiza parvula.
4. Certhidea olivasea.

FIGURE 1.4. A drawing by the ornithologist John Gould in Darwin's *The Voyage of the Beagle*, showing several of the finches that are now often referred to as "Darwin's finches."

of Species. The most striking passage describes the finches of the Galápagos Islands: "Seeing this gradation and diversity of structure [of bills] in one small, intimately related group of birds, one might really fancy that from an original paucity of birds in this archipelago, one species had been taken and modified for different ends" (figure 1.4). This was his first published hint of what he had been thinking.

Birds figure in eleven of the fourteen chapters of *The Origin of Species*. Darwin had taken up pigeon breeding in order to learn how the diverse varieties of domestic pigeons had been formed by human selection of native Rock Doves (*Columba livia*). Pigeons dominate chapter 1 and recur throughout the book.[15] He describes how novel, inherited variations gave rise to breeds that differ more than whole genera and families of birds: "the Jacobin has the feathers so much reversed along the back of the neck that they form a hood"; the fantail has "thirty or even forty tail-feathers, instead of twelve or fourteen—the normal number in all the members of the great pigeon family" (and in many other bird families as well). In developing the idea of a "struggle for existence," he points out that the (Northern) Fulmar (*Fulmarus glacialis*) lays only one egg but is "the most numerous bird in the world"—showing that most species, which lay more eggs but are rarer, must suffer immense mortality. Writers on "natural theology" cited species' adaptations as evidence of the wisdom and beneficence of the Creator, but Darwin saw that many species have features that serve no function and instead show evidence of their ancestry. Referring to the Campo Flicker (*Colaptes campestris*; plate 3), he writes that "on the plains of La Plata, where not a tree grows, there is a woodpecker, which in every essential part of its organisation, even in its colouring, in the harsh tone of its voice, and undulatory flight, told me plainly of its close blood-relationship to our common species; yet it is a woodpecker which never climbs a tree!" Likewise, "what can be plainer," he asks, "than that the webbed feet of ducks and geese are formed for swimming? Yet there are upland geese with webbed feet which rarely go near the water," which he had seen in Chile.

In pondering human evolution, Darwin ventured that some human features could best be explained by what he called sexual selection, the subject of more than half of *The Descent of Man, and Selection in Relation to Sex* (1871). Sexual selection, meaning variation in mating success, was Darwin's explanation for many of the colors, crests, exaggerated tail feathers, wattles, behavioral displays, and other features (especially of males) that delight and intrigue everyone who has any interest in birds. Drawing on both his own observations and natural history literature, Darwin devotes four full chapters to birds and cites at least 170 species. Birds provided more evidence for his ideas about sexual selection than any other group of animals (figure 1.5).

I'll mention just a few later ornithological contributions to evolutionary science. Let's start in 1899. Darwin had presented massive evidence for descent with modification—the fact of evolution—but no direct evidence of natural selection. Apparently, neither he nor anyone else, for almost forty years after *The Origin of Species*, thought it would be possible to detect and measure natural selection, which Darwin thought would work extremely slowly, like the uplift of mountain ranges. Hermon Bumpus, in 1899, was one of the first to report natural selection. He made measurements of 136 distressed House Sparrows that were collected near Woods Hole, Massachusetts, after a severe winter storm. About half died and half recovered. Bumpus reported that those that deviated most from the average succumbed,[16] a pattern that we now call "stabilizing selection," which will maintain the status quo. Since then, similar measurements on hundreds of species have shown natural selection on many characteristics.

David Lack, whose career was mostly at Oxford University, christened the finches in the Galápagos Islands "Darwin's finches" in the 1940s and was the first to propose that they formed an "adaptive radiation," in which the related species had become adapted by natural selection to specialize on different food items. He suggested that they illustrated Darwin's idea that species diverge—become different—because specialization enables them to escape competition with other species.[17] This principle later became a major focus of ecologists who wanted to understand how species can coexist. Lack made an even more important contribution when he proposed and verified an explanation for why birds lay a certain number of eggs and no more. This was the start of a whole field of study: understanding why species have evolved differences in their reproductive rate and other aspects of their life history. I'll describe Lack's study in chapter 7.

FIGURE 1.5. A South American Long-wattled Umbrellabird (*Cephalopterus penduliger*), in Darwin's *The Descent of Man, and Selection in Relation to Sex*. The male's wattle is one of many examples that Darwin proposed had evolved by sexual selection.

Modern thinking about *speciation*—the process by which a species splits into two or more descendant species—starts with Ernst Mayr's studies of birds. Mayr entered biology through bird-watching as a young student in Germany. From 1927 to 1930, he collected birds in New Guinea and the Solomon Islands, and in 1931 he was hired by the American Museum of Natural History in New York to curate its large collection of birds from that region. These were the basis of his many papers on bird taxonomy and of his ideas about speciation. He presented these ideas, and synthesized a vast amount of evidence about what species are and how they are formed, in his 1942 book *Systematics and the Origin of Species*.[18] He continued to publish prolifically on evolution, bird systematics, and the philosophy and history of biology until his death in 2005 at the age of one hundred.

The family tree, or phylogeny, of species is both a basis for classification and a key to understanding the evolutionary history of species and their characteristics. Determining relationships among species is not easy, especially when it is based on similarity in just a few anatomical features. The muscles that operate a bird's sound-producing structure (syrinx) helped to distinguish perching birds (Passeriformes) as a distinct evolutionary branch, but it was hard to be sure that this one characteristic was a reliable index of relationship.

The development of modern, more reliable methods came with the use of molecular information, especially DNA. Many evolutionary biologists developed (and continue to improve) these methods, but the first person to apply them to a large group of organisms was one of my teachers, the ornithologist Charles Sibley at Cornell University (later at Yale). The best DNA technology in his day, before DNA sequencing on a large scale was feasible, was a crude method, "DNA–DNA hybridization," which Sibley and his long-term collaborator, Jon Ahlquist, used to estimate relationships among thousands of species of birds.[19] Many of Sibley's claims were certainly wrong—but some were both surprising and apparently correct.

Two founders of the science of animal behavior, Konrad Lorenz in Germany and Niko Tinbergen in the Netherlands, developed their ideas partly by studying birds. Tinbergen, whose books included *The Herring Gull's World*, emphasized that one of the most important questions to ask of any animal's behavior is how and why it evolved. This approach has been a dominant theme in animal behavior since the 1960s, led largely by researchers on bird behavior. Similarly, David Lack, Robert MacArthur, Jared Diamond, and other students of birds were among the scientists who shaped an evolutionary approach to ecology, the scientific study of interactions between organisms and their environment.

These are some of the pioneers who drew on birds to develop major areas and ideas in evolutionary biology. A great many other students of birds, then and since, have developed these subject areas and have gained insight into the evolutionary history and causes of birds' characteristics and diversity. Some readers will have noticed that all the biologists I have mentioned so far are (White) men—as in most of biology and other sciences. It was only in 1937 that one of the first major studies by a woman was published: "Studies in the life history of the song sparrow," by Margaret Morse Nice[20]—a study more ecological than evolutionary in nature. Happily, women have more recently contributed many pathbreaking studies, as I describe in later chapters. Biologists throughout the world, including Asia and Latin America, conduct research on bird evolution, but unfortunately fewer people of color in the United States and Europe do so (although this too is changing).

So birds have been major players in all kinds of evolutionary studies—and this research has cast light on almost any aspect of bird biology and diversity you can think of. In the following chapters, I begin with evolutionary relationships among birds and what the fossil record tells of their history (chapters 2 and 3). I then turn to the process of adaptation by natural selection as revealed by studies of birds (chapter 4), to variation within bird populations—the raw material of evolution (chapter 5)—and to a few fascinating examples of birds' adaptations and how they are studied (chapter 6). The next three chapters answer questions about how birds' diverse life histories (chapter 7), sex lives (chapter 8), and social behaviors (chapter 9) have evolved by natural selection. After asking what species are and how they evolve (chapter 10), I return to birds' evolutionary history, asking what accounts for the geographic distribution and diversity of different groups of birds (chapter 11). Finally, in chapter 12, I ask what light evolutionary studies might cast on the future of birds: their survival or extinction in a world reshaped by humans.[21]

2

Parrots, Falcons, and Songbirds

THE BIRD TREE OF LIFE

As a young birder in the 1950s, I started to learn bird taxonomy. The Red Knot was *Calidris canutus*, the Sanderling *Crocethia alba*, the Least Sandpiper *Erolia minutilla*. The Pine Siskin (*Spinus pinus*), Rose-breasted Grosbeak (*Pheucticus ludovicianus*), and White-throated Sparrow (*Zonotrichia albicollis*) were all in the family Fringillidae. Today, all those sandpipers are combined with many others in the genus *Calidris*, to express the conviction that they all come from a common ancestor and are more closely related to one another than species in other genera. The siskin is still in Fringillidae, but the grosbeak and the sparrow are in different, newly named families. It is now understood that these three groups of birds, despite their similarities, are not as closely related to each other as they are to some other bird groups. Ornithologists change bird classification because they want it to reflect their understanding of the relationships among bird species, and that understanding is growing all the time—faster now than ever before. This chapter is about how ornithologists do that and what we can learn about birds as a result.

The concept that species are actually related to each other by descent—that all bird species, for instance, stem from one ancestral species—was one of Darwin's most profoundly important ideas. The only illustration in *The Origin of Species* is a diagram with horizontal lines, like geological strata, representing different times in the past, and vertical lines that represent species persisting through time (although some end in midpage: they become extinct). Most of the species lineages split into two branches—two species—each of which may later split again. Thus a number of species trace their ancestry back to a specific ancestral species (their common ancestor). In the spirit of Darwin's drawing, here is one such branching lineage, with time going from left to right instead of upward, so that the tree "grows" from its "root" at the left.

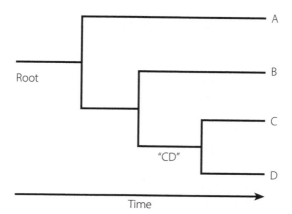

Darwin's diagram (now called a phylogenetic tree, or simply a phylogeny) conveys his revolutionary idea that species are related by common ancestry. The more recent the common ancestor of any two species, the more closely related they are—just as you are more closely related to your sibling (through your parents) than to your cousin (through grandparents). So species C in the diagram is more closely related to D than to B or A, and B, C, and D are more closely related to each other than to A. The tree conveys the relative sequence of branching that gave rise to the four species. A line segment between two branch points (such as "CD" in the figure) is an ancestral species lineage. Darwin wrote that when his ideas become accepted, "our classifications will come to be, as far as they can be so made, genealogies, and will then truly give what may be called the plan of creation." That is what bird taxonomists are trying to do. (Darwin's use of "creation" doesn't imply religion or the supernatural.)

Two species[1] from an exclusive common ancestor (such as C and D) are called *sister species* (and two species-groups from a single ancestor are "*sister groups*"). They will be alike in many characteristics, which they have inherited from their common ancestor, but they differ in whatever characteristics (and genes) have evolved in both of them since the split. Often, more closely related species are more similar to each other than to other species, but not always. Sometimes, one species (say, C) will undergo a lot more evolutionary change than its relative (D), so there may be a weak correspondence between relationship and similarity in some features. (You are more closely related to your sister than your cousin, even if you and your cousin are more similar in hair and eye color. Relationship and similarity are not the same.) When I saw the Ground Tit (formerly *Pseudopodoces humilis*, figure 2.1) on the Tibetan Plateau, I thought it hopped awkwardly, almost ludicrously, on the treeless plain. It was formerly considered a ground jay, and its shape and long bill make it look utterly unlike any familiar tit or chickadee. However, it turns out to be closely related to the Great Tit (*Parus major*) and is now named *Parus humilis*.[2] Compared with the Great Tit, it has evolved more, and therefore more rapidly, since their common ancestor: rapid evolutionary change can make relatives look unrelated. On the other hand, similarity can mislead us into thinking species are closer relatives than they actually are. Crows, grackles, and the European Blackbird (*Turdus merula*) have much the same color, but we know (from other characteristics) that they are related to species in three different families. Their black color is an example of *convergent evolution*—independent origin of similar (or even identical) features in different lineages. So birds' obvious characteristics aren't always strong indicators of relationship and may not be useful to classify them.

Most taxonomists hold that the best evidence of relationships, when anatomical characteristics are used, is "shared derived" characteristics that have evolved uniquely in the common ancestor of certain species to the exclusion of other species. ("Derived," meaning newer, is contrasted with "ancestral," the older condition.) For example, among living animals, only birds have feathers, which have such complex structure that we are confident they evolved only once and are evidence that birds are all more closely related to one another than to any other living creature. Birds share many anatomical features, such as bipedal posture and a reversed hind toe, only (or almost only) with theropod dinosaurs. The theropods, including the birds, form a single branch (or *clade*, which is simply Greek for branch) of the phylogenetic tree of life (see chapter 3). Modern birds share several derived features, such as fusion of several tail vertebrae into a single bone (the pygostyle). Various other features that are thought to have evolved only once mark the common ancestry of some living birds, such as the songbirds (order Passeriformes), which have very distinctive sperm cells, with a spiral head and a membrane along the tail, that swim by spinning rather than by beating the tail.[3] Many species of Passeriformes are classified as a suborder, the oscines, because they share a complex set of muscles in the syrinx ("voice box"). On the premise that these

FIGURE 2.1. The Ground Tit (*Parus humilis*; right) is closely related to typical tits, such as the Great Tit (*Parus major*; left), but has very different features owing to rapid adaptation to living on the ground in a treeless habitat. (Art, Luci Betti-Nash.)

several features evolved only once, we can draw a phylogenetic tree that portrays our hypothesis about relationships among some birds, dinosaurs, and more distantly related reptiles such as crocodilians.

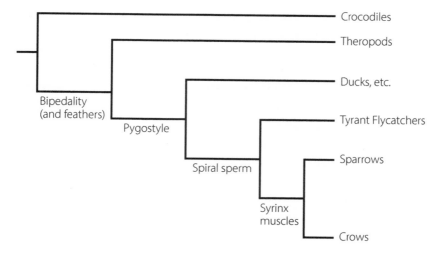

The phylogeny portrays groups that evolved within other groups, each distinguished by features that we suppose evolved only once. Crows, flycatchers, and other Passeriformes are descended from an ancestral lineage that evolved spiral sperm; this lineage was one of many descendants of a lineage that evolved the pygostyle; and this, in turn, was one of many descendants of the lineage that evolved bipedality, reversed first toe, feathers, and other features of theropod dinosaurs.

A problem in determining relationships among various living birds is that rather few of their characteristics have unequivocally evolved only once. Taxonomists learned long ago that you need multiple lines of evidence—multiple features, the more the better—to be confident. That is exactly what is provided by DNA sequences, which have become the most important data, by far, for determining phylogenies (of all kinds of organisms, from bacteria and viruses to birds and humans). At this point, we have to be clear on some of the basics of DNA; they will come up in many contexts in this book. All we need to know at this point can be summarized in four paragraphs.

Each chromosome in a cell's nucleus is a rod-shaped structure made up of some proteins and an extremely long DNA molecule, parts of which encode proteins and are called genes (figure 2.2). DNA consists of two matched strands, each of which is a sequence of small molecules (bases, or nucleotides), of four types: A, T, C, and G (adenine, thymine, cytosine, guanine). The other strand has complementary bases: A and T always pair, as do C and G. So if one strand reads TCGGAC, the other reads AGCCTG. If we refer to the sequence of one strand, we don't need to specify the completely predictable sequence of the other strand. Every time a cell divides into two daughter cells (which might be sperm or eggs), the two strands separate, and bases are brought in to pair with the exposed bases on each strand. If a base pair in the double-stranded DNA is T-A, an A joins with the exposed T on one of the separated strands, and a T joins with the exposed A on the other. Then the two half-new chromosomes separate into the two new daughter cells. This process of DNA replication is astonishingly faithful, but mistakes are occasionally made, resulting in a new base pair. This is the most frequent kind of *mutation*, but any one base pair mutates only about once or twice per billion cell divisions.[4]

In most genes, one of the strands is "read" by complex molecular machinery to be a code for the twenty kinds of small molecules (amino acids) that are most often strung together to form a protein. (The sequence of amino acids determines how the protein folds itself into a 3-D configuration, which determines how the protein functions.) The amino acids are encoded by triplets (codons) of successive bases: for instance, CGA encodes alanine (figure 2.2). There are sixty-four triplet combinations (four possible bases at each of three

(A) DNA replication

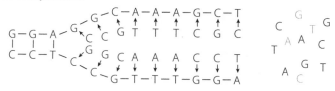

(B) Transcription and translation into protein

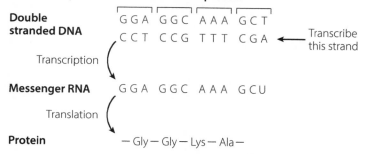

FIGURE 2.2. DNA, RNA, and protein. (A) A short DNA sequence that is partly separated as it is replicated. Replication is proceeding from right to left. Nucleotide bases A, T, C, and G (from a pool, at right) bind to their partners (A with T, etc.) on each of the separate strands. If a wrong base binds, that can lead to a mutation. (B) Four triplets (codons) of base pairs of double-stranded DNA are shown. The lower strand is the template for synthesis of a single-stranded messenger RNA, with bases complementary to those in the transcribed strand (e.g., G in RNA where DNA has C; RNA has U instead of T). Each RNA triplet specifies an amino acid that is added to a growing protein chain. For example, GGA specifies Gly (glycine). GGA and GGC are synonymous: both specify glycine, so mutation of the corresponding DNA triplet CCT to CCG would be a synonymous mutation.

positions), so some codons encode the same amino acid. Both CGA and CGG, differing only in the third base, encode alanine. But change CGA to AGA, and you get threonine. A mutation that doesn't change the amino acid is *synonymous*; if it does change the amino acid, it is *nonsynonymous*. The difference between nonsynonymous and synonymous mutations is really important because synonymous mutations don't alter the protein or its function and therefore have virtually no effect on the organism, whereas nonsynonymous mutations may have effects. Synonymous mutations neither help nor harm: they are "selectively *neutral*" and can increase and become fixed by genetic drift (cf. figure 1.3 in chapter 1).

Although a small fraction of nonsynonymous mutations are advantageous or neutral, many are harmful: they impair protein function, often affect physiological or morphological characteristcs, and reduce individuals' survival or reproduction. For this reason, they tend not to persist in species populations, and so genes evolve more slowly by nonsynonymous mutations than by synonymous substitutions. The genome also has a lot of noncoding DNA, some of which appears to be nonfunctional and which also evolves rapidly, like synonymous mutations, by genetic drift. These differences are important for phylogenetic studies because rapidly evolving parts of the genome provide information on relationships among species that diverged over a short time interval in the recent past. (I mean "short" and "recent" by geological standards, namely hundreds of thousands of years!).

Most of an animal's genes are on the chromosomes, and most are represented by two copies, one from each parent, which may differ slightly in base pair sequence because of past mutations. These are called nuclear genes. A few genes are in mitochondria: structures in the cell that proliferate like bacteria (because they descended from symbiotic bacteria, more than 1.9 billion years ago). Mitochondria are inherited only from the mother. In the early days of DNA-based phylogenetic studies, most researchers used mitochondrial DNA because it was easier to sequence for several technical reasons.

Because genes are replicated and pass from one generation to the next, we can think of a "parent" gene copy and its "descendant" genes: a gene lineage that is carried on from generation to generation. Every gene in my cells, or of a House Sparrow, has descended from a DNA sequence that existed more than 4 billion years ago, in the earliest forms of life. So many mutations may have accumulated in the sequence that it has become unrecognizably different from that ancestral DNA; also, animals have far more genes than the earliest life forms did because of a process that duplicates genes within genomes. No wonder Richard Dawkins (in *The Selfish Gene*) called DNA the "immortal coil."[5] That all organisms' genes are actually descended from ancient common ancestors isn't a fanciful idea: some of the same genes program the development of eyes in birds and humans, and even in insects. Biologists replaced 414 genes that are necessary for yeast to survive with recognizably similar human genes. Almost half of the human replacements enabled the yeast to survive![6]

Because genes have descendants, we can imagine a genealogy, or "phylogenetic tree," of a gene and its descendant gene copies—a *gene tree*—just as Darwin imagined a *species tree*. Now suppose that in a sperm cell of an individual goose, a mutation happens in base pair number 130 in a 1,000-base pair-long DNA sequence: G changes to A (and on the other strand, C to T, but we don't need to say that). One of the goose's offspring inherits this mutated gene, and the inheritance continues for thousands of subsequent generations. All the gene copies descended from that originally mutated gene, carried by some number of geese, will have the mutation and form a branch of "related" genes, more closely related to each other than to other gene copies that don't have the mutation (assuming the same mutation doesn't originate repeatedly). These gene copies might be distributed among several species of geese, if they have evolved from the ancestral species in which the mutation first occurred.

Now we can return to determing relationships among species—using relationships among their genes. Suppose we aim to determine the relationships among Brant (or Brent Goose, *Branta bernicla*), Canada Goose (*Branta canadensis*), and Nene (or Hawaiian Goose, *Branta sandvicensis*). We sample a number of individuals of each species, and we find that at site 130 in a certain gene individuals of the same species are identical, but that Canada and Nene have an A and Brant has G. We might suppose that Canada and Nene are closest relatives and that an original G mutated to A in the common ancestor of these species, as shown in tree 1.

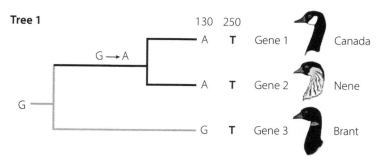

But here is an alternative possibility: the common ancestor of all three species had A, which mutated to G in the Brant, which is actually related to Nene.

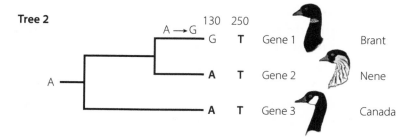

How might we tell which is more likely? By scoring additional species that we are sure are more distantly related to these three than these are to each other. We refer to these three (the species whose relationships we want to know) as the "ingroup" and other species, which we are confident are more distantly related, as "outgroups." We are quite sure that lack of feathers is the ancestral condition, and possessing feathers is the derived (more recently evolved) condition because all the more distantly related reptiles—crocodilians, lizards, in fact all other vertebrates—lack feathers. Similarly, if we find that site 130 in this gene is G in many other genera of geese (outgroups), the simplest inference is that having A at this position represents an evolutionary change from G during the history of the ingroup. Let's choose Emperor Goose (*Anser canagicus*) and Cape Barren Goose (*Cereopsis novae-hollandiae*). Ornithologists are quite confident that the *Branta* species are more closely related to each other than to the Emperor or Cape Barren because of structural and other differences among these genera. Suppose we find that they both have G at position 130.

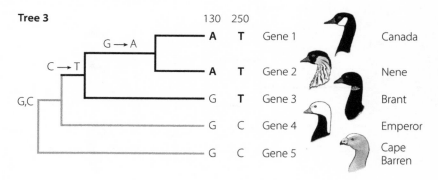

Tree 3

This increases our confidence that G was present in the common ancestor of the species of *Branta* and was replaced by A in the common ancestor of Canada and Nene. If so, these are closest relatives. I have supposed further, in creating this diagram, that the ingroup species share T at another position (250), in contrast to the outgroups. An observation like that would increase our confidence that the three *Branta* species are really more closely related to each other than to Emperor and Cape Barren geese.

In deducing that Canada and Nene are closest relatives, we have made a big assumption: that each mutational change happened only once in this history. Let us look at some real data. Not coincidentally, in 1991 Thomas Quinn, Gerald Shields, and Allan Wilson published a study of the Emperor and the

TABLE 2.1. Corresponding DNA Nucleotides in a Mitochondrial Gene in Four Species of Geese

	Base pair						
	3	4	9	12	15	18	21
Canada	ACC	GCC	GTA	CTT	CTG	CTA	GAT
Nene	ACC	GCC	GTA	CTT	CTG	CTG	GAC
Brant	ACT	GCC	GTC	CTC	CTA	CTA	GAT
Emperor	ACT	TCC	GTC	CTC	CTA	CTA	GAC

Branta species based on DNA sequences of the mitochondrial gene cytochrome *b*.[7] Wilson, a New Zealander who became a professor at the University of California at Berkeley, was a creative, pioneering leader in using DNA for phylogenetic studies. In table 2.1 are just a few of the nucleotide triplets from the very long gene sequence for the four species. Most triplets were identical in all the species.

The base pair that I have numbered 4 differs between Emperor and all the *Branta* species and is consistent with the three *Branta* species being more closely related. At bases 3, 9, 12, and 15, Brant has the same base as Emperor— probably an ancestral condition, inherited from the common ancestor of all four species. Canada and Nene share a mutation at each of these sites: the strongest evidence that among the three ingroup species, Canada and Nene are the closest relatives (as in tree 1). This is indeed the conclusion that Quinn and his coauthors drew.

But look at base 21, at the right end. Emperor shares C with Nene, not with Brant as in the other cases. Brant and Canada have a T at this site, so might Canada and Brant be closest relatives, with C mutating to T in their common ancestor? That would be a phylogeny contrary to the one implied by bases 3, 9, 12, and 15.

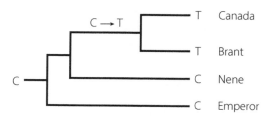

The alternative is that Canada and Nene really are closest, but our assumption that a base mutated only once is wrong.

Tree 2

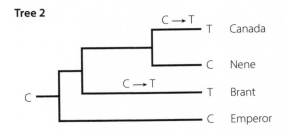

In other words, there was convergent evolution of T from C twice in the history of these species. Just as shared black color might lead us wrongly to suppose that crows are more closely related to blackbirds than to jays, certain changes in DNA bases can lead us astray because of convergent evolution (or reverse changes of base pairs). The important message is that whether we use anatomy or DNA, different characters can imply different phylogenetic relationships. So researchers must use many characters. Nowadays, this means using many varying base pairs in many genes (hence, total sequences of many thousands of base pairs). If one phylogeny is favored over another by a substantial number of characters, that becomes the best estimate of the history of relationships. In some cases, the preponderance of evidence for a particular phylogeny is so great that we can be very confident in it. A later study of goose phylogeny, based on DNA sequences from more than 5,000 genes, produced the same relationships among these four species that Quinn and his colleagues had proposed.[8]

Summing up: we take corresponding DNA sequences from the species whose relationships we want to know (ingroup) and some more distantly related species (outgroup). We assume, provisionally, that any base pair shared by the outgroup and one or more ingroup species is the original (ancestral) state. A mutated variant of that base pair that is found in two or more ingroup species is a shared change that may have happened in their common ancestor, giving evidence that these species are closely related. Some base pairs, by the same logic, might imply different relationships among the ingroup species. That is, different characteristics disagree. But if many more DNA sites point to one relationship over another, we assume that the majority vote wins. Among the possible phylogenetic configurations, the one that is supported by the greatest number of characteristics (DNA bases) is our best estimate of relationships. It is not necessarily the absolute truth—in fact, no careful scientist, in any field, ever claims absolute truth or certainty. It is, instead, our best current understanding—in which we might have more or less confidence, depending perhaps on how much data we have. It may also depend on how powerful and appropriate the methods are that we used to analyze the data. The approach I've described is called the "parsimony method," which can be the best method for certain data but by itself has some drawbacks. It is an ingredient of more complicated statistical methods of phylogenetic analysis that have become the norm today and which use massive DNA sequences and take account of processes of DNA sequence evolution.

(As an aside: the Covid-19 pandemic broke out during the preparation of this book. Epidemiologists determined that the coronavirus spread to the western United States from China but that the East was most immediately infected from Europe. To trace the spread, they used the phylogenetic methodology that evolutionary biologists developed to determine species trees and gene trees. Basic research can have unforeseen importance!)

With all the pitfalls that beset ornithologists who try to reconstruct the history of bird evolution, it is amazing how many bird relationships that had been founded on traditional morphological features have passed the test of DNA analysis. Earlier, I described how features such as the muscles of the syrinx and some other features provided evidence of the common ancestry of the songbirds or perching birds (order Passeriformes) and, within them, the division into suboscines (tyrant flycatchers, antbirds, pittas, etc.) and oscines (most other familiar songbirds). Several research groups have published phylogenetic trees of all the major groups of birds based on a huge amount of DNA sequence data.[9] In all these studies, the various species classified as passeriforms came out as a single clade: they indeed all seem to share a more recent common ancestor with one another than with any other bird. Similarly, these studies agreed with the grouping into suboscines and oscines (figure 2.3). The DNA analyses confirmed morphology-based suggestions that the New Zealand wrens (Acanthisittidae) are the sister group to all other passeriforms, the Australian lyrebirds are the sister group to all the other oscines, and many other aspects of relationships that generations of ornithologists proposed before DNA could be sequenced—or even before DNA was known to be the genetic material! When very different sources of information converge on the same answer, we gain confidence that the answer is right.

DNA studies have solved some long-standing puzzles—at least tentatively. What are flamingos related to? Some suggested ducks, others ibises, but DNA evidence points to grebes—as unlikely at that seems! In some cases, the new evidence overturns some long-held views. Some ornithologists had suspected

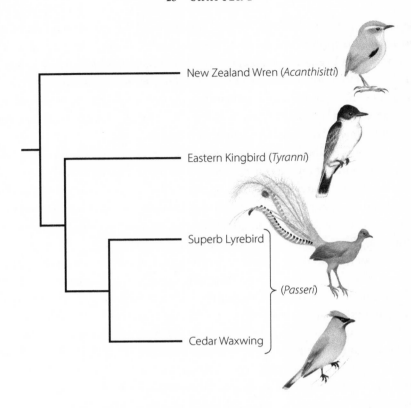

FIGURE 2.3. A phylogeny of some major lineages of the perching birds, order Passeriformes. Top to bottom: New Zealand Rockwren (*Xenicus gilviventris*, suborder Acanthisitti), Eastern Kingbird (*Tyrannus tyrannus*, suborder Tyranni, suboscines), Superb Lyrebird (*Menura novaehollandiae*), Cedar Waxwing (*Bombycilla cedrorum*) (both in suborder Passeri, oscines). (Art, Luci Betti-Nash.)

that the Falconidae (falcons and caracaras) were convergent with, not closely related to, the hawks and eagles (Accipitridae), but I don't think anyone had suspected today's consensus that they are the sister group of parrots plus songbirds. The tanager family, Thraupidae, has lost the genus most familiar to North American birders (*Piranga*, e.g., Scarlet Tanager, *P. olivacea*) but has gained seedeaters, yellow-finches, and South American cardinals, formerly placed in different families but recently found to be intermingled with the branches of traditional tanagers in the thraupid phylogeny.[10]

So DNA will soon answer all our questions, we will know the full family tree of all birds, taxonomists will use this to create a classification that will be stable, all the checklists and bird guides will settle on a list order, and birders can breathe a sigh of relief.

Maybe, but that day won't come for quite a long while. There are several ways, well understood by evolutionary biologists, in which evolution can make some phylogenetic relationships very hard to disentangle.[11]

One is hybridization. Closely related species sometimes hybridize at a low frequency, so some genes that developed sequence differences in one species may become prevalent in a related species. For instance, different genes implied different relationships among the Sharp-tailed Grouse (*Tympanuchus phasianellus*), the Greater Prairie Chicken (*T. cupido*), and the Lesser Prairie Chicken (*T. pallidicinctus*), which are known to hybridize occasionally.[12]

Another is time. Relationships can be very hard to determine if three or more species (say X, Y, and Z) all originated over a very short time from their common ancestor (A) because very few evolutionary changes occur in the brief time between successive branching events. Suppose ancestor A gave rise to the species "XY" and the species Z, and then "XY" divided into today's species X and Y. If the time interval between those splits (speciation events) is very short, there will be very few base pair replacements, and species X and Y will share few distinctive DNA markers, to the exclusion of Z, as evidence that they are the most closely related species. We may not be able to determine confidently the relationships among the species. In some cases, many species originate within a short time. The problem is the same, whether the burst of diversity occurred recently or very long ago.

For example, Irby Lovette, now at the Cornell Laboratory of Ornithology, used mitochondrial genes to estimate the relationships among twenty-four species of wood warblers in the genus *Setophaga* (formerly *Dendroica*, the name when this study was published).[13] These include many brilliantly colored species, such as the Blackburnian (*S. fusca*), Cerulean (*S. cerulea*), and Townsend's Warblers (*S. townsendi*) that many American birders eagerly seek in spring (figure 2.4). Using the phylogeny, Lovette estimated the percentage of DNA base pairs that had changed between each pair of species since their common ancestor. About half of the forks in the phylogenetic tree—about eleven lineages—fall within a narrow range of sequence divergence, so the number of derived base pair states that any two species share isn't very different from the number each shares with some of the other species. As a general rule, it takes longer to evolve a big DNA sequence difference than a small one; in fact, ornithologists use a very rough "molecular clock," according to which mitochondria DNA evolves at about 2% change per million years. (I return

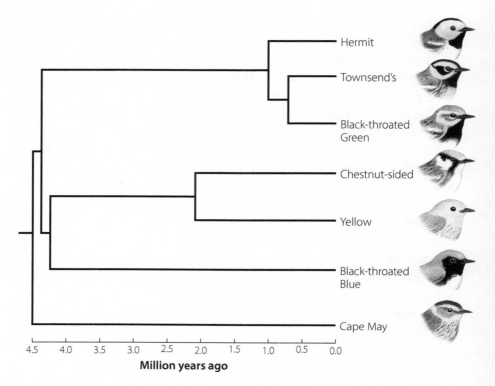

FIGURE 2.4. A partial phylogeny of the warbler genus *Setophaga*. (After Lovette and Berming-ham 1999; art, Luci Betti-Nash.)

to molecular clocks in chapter 3.) Lovette concluded that the diversification of these warblers began with a great burst of speciation, so fast that he could not be confident about just how these eleven lineages are related to one another.

Rapid diversification in a short time period is one of the most common causes of uncertainty about relationships. As we'll see in the next chapter, many of the major lineages of birds (orders and groups of orders) originated during a relatively short geological time span, and their relationships are uncertain despite extensive data and the efforts of capable scientists.

So as is true for science generally, phylogenetic studies can produce uncertain or wrong conclusions—especially if they are based on insufficient data. When you hear that "they" have "shown" new and unexpected relationships among some bird species, you might sometimes want to wonder how strong the evidence is and wait a few years to see if "they" come up with a different story. That's how science works: maintain a little skepticism, consider every conclusion as slightly tentative, gain more confidence as further studies agree.

But a healthy skepticism doesn't relegate all results to the rubbish bin. As I described earlier, many aspects of bird relationships, even based on traditional anatomical features, have stood the test of time. And many new proposals have been supported by multiple studies, by different researchers using different data. The affinity of the falcons and parrots to the songbirds is one of many examples. It is a truly exciting time to be a student of evolution, whether of birds or anything else, because we now have phenomenal tools for answering questions with unprecedented rigor on an unprecedented scale! So, let's accept that some bird relationships are well understood, or are on their way, and ask how such information can be used.

Especially in concert with molecular clocks, phylogenies often provide evidence on timing: when lineages split and when various evolutionary changes happened. Chapter 3 and some later chapters include interesting examples. Here I touch on two other uses of phylogenies: piecing together the evolutionary history of birds' characteristics and classification.

For biologists, one of the ways phylogenies are most useful is to reconstruct the history of evolution of interesting characteristics. The phylogenetic tree of all major bird groups (and of extinct bird relatives) implies that all living birds came from a flying ancestor. Flightless species must have evolved from flying ancestors. Flightless birds include the ratites (mostly very large birds such as ostriches and cassowaries), Fuegian Steamer Duck (*Tachyeres pteneres*), Dodo (*Raphus cucullatus*, in the pigeon family), penguins (family Spheniscidae), the Takahe (*Porphyrio hochstetteri*, a large flightless gallinule, plate 4), and the Kakapo (*Strigops habroptila*, a huge, peculiar New Zealand parrot, plate 5). Each of these is clearly related to flying birds (the Steamer Duck to other ducks, the Kakapo to other parrots, etc.), so flightlessness must have evolved independently in each case. The Great Tit and the Ground Tit (figure 2.1) also show how we can retrieve history from phylogenies. The Great Tit resembles other members of the tit family in most respects, while the Ground Tit differs greatly in its habitat and structure. Since the Ground Tit is one twig within the larger tit family tree, its distinctive features must have evolved from those of a standard-issue tit, rather than vice versa. Likewise, recall that we postulated that a DNA base pair is ancestral relative to another base pair at that site if it is shared between outgroup species and certain ingroup species. That is a statement about the history of DNA sequence evolution. Based on phylogeny, biologists often infer ancestral DNA sequences, synthesize and test the encoded proteins, and infer the biochemical capabilities—such as color vision—of extinct ancestors! I will describe an example in chapter 6.

How does phylogenetic research affect bird classification? Most changes of classification are of two kinds. One is at the species level: splitting one species into more than one or combining (lumping) two into one. A major issue here is, "What is a species?" That is one of the subjects of chapter 10. The other change is determining what the "higher taxa" of birds will be and to which higher taxon a species belongs. A *taxon* is a named category of classification, such as the genus *Passerina*, the family Cardinalidae, the order Passeriformes. "Higher" taxa are those above the species level: genera, families, orders, and various sub- and supercategories. Orders are divided into families, families into subfamilies (sometimes), these into genera, and genera into species. (Incidentally, the ending -formes signifies an order, -idae a family, and -inae a subfamily.)

Recall Darwin's prophecy that "our classifications will come to be, as far as they can be so made, genealogies." Most taxonomists have endeavored to create classifications that reflect relationships to a large extent, meaning that the species in a higher taxon share a common ancestor. Thus all the members of the class Aves, namely all the birds, are thought to stem from a single species of "protobird" in the distant past. The order Falconiformes included families such as Accipitridae (hawks and eagles), Cathartidae (New World vultures), and Falconidae (falcons), each presumed to have descended from a single ancestral species that had descended from the original ancestor of the entire order. Phylogenetic analysis with DNA has shown that the Falconidae are not closely related to the other families, which are now referred to as order Accipitriformes.

Modern taxonomists tend to differ from traditional ones on one aspect of what it means for a classification to reflect evolutionary ancestry. Traditionalists sometimes split out certain members of an evolutionary branch into a separate taxon to call attention to particularly striking evolutionary divergence. The traditional class Reptilia included the familiar reptiles as well as dinosaurs—but not the birds, which were placed in the separate class Aves even though everyone knew that they are related to dinosaurs and crocodiles. They had class—they were privileged because of their very different lifestyle and features. So all the members of Reptilia were derived from a common ancestor—but not all the descendants of that ancestor were included in

Reptilia. The Reptilia were incomplete ("paraphyletic," to use the modern jargon). Modern taxonomists will have none of this halfway measure. A taxon should be "monophyletic," they insist: it should include *all* the species that stem from a common ancestor. So birds should be in Reptilia along with crocodiles (their closest living relatives).

This taxonomic "philosophy" has wrought many changes in bird classification. Here is an example that will resonate with many American birders, who have long used the genus name *Dendroica* when referring to a large number of warbler species. Earlier, I described Irby Lovette's phylogeny of this group. Lovette included a number of other species of warblers in his analysis.[14] He found that according to the mitochondrial DNA, the American Redstart (*Setophaga ruticilla*) is inside the *Dendroica* cluster: it appears to be more closely related to a clade of nineteen species of *Dendroica* than are four other *Dendroica* species. Adhering to the principle that a taxon such as a genus should be monophyletic, Lovette had two choices. He could keep the name *Dendroica* for the clade of nineteen species, keep the American Redstart as the only member of *Setophaga,* and bestow four new genus names on the four distantly related dendroicas. Or he could "sink" the name *Dendroica* altogether and rename the group of twenty species *Setophaga.* That is what he did. Why call them all *Setophaga* instead of calling the American Redstart a *Dendroica*? Because one of the most important rules in the International Code of Zoological Nomenclature (yes, there is a code, and it is "enforced" by a panel of experts who hear appeals!) is the "Law of Priority." It states, "The valid name of a taxon is the oldest available name applied to it," and that "a family-group taxon formed by the union of two or more taxa of that group takes the oldest valid family-group name among those of its components." So because the American Redstart was described under the name *Setophaga* in 1827, before the first warbler was named *Dendroica* (in 1842), they all have to be called *Setophaga* when they are combined into a single genus.

A rose by any other name would smell as sweet, but even though names are conventions, we are in error if we misuse them: I would be wrong to call a rose a skunk cabbage. So changes in classification are often required by new understanding of species' relationships. For example, a species may turn out to be in the wrong place: it doesn't share a recent common ancestor with the other members of the taxon. The Scarlet Tanager and other members of its genus are not closely related to other members of the large tanager family (Thraupidae) but instead are among the cardinal grosbeaks (Cardinalidae). Sometimes a higher taxon turns out to include distantly related groups, each more closely related to something else, as the old Falconiformes proved to be.

Suppose the phylogeny of a big group of species has been confidently worked out. What determines whether the various branches are designated as genera, or subfamilies, or families—their taxonomic "rank"? This is pretty arbitrary; everyone agrees that (except for the principle of monophyletic groups) there are no objective criteria. It usually comes down to the (subjective) degree of difference. Ornithologists have known for a long time that the tropical American toucans, designated as family Ramphastidae, are closely related to the barbets, classified as the family Capitonidae and occuring in the Neotropics (tropical America), Africa, and Asia. Although toucans and barbets are very similar, the toucans were classified as a separate family, mostly because of their much larger bill. Today, we know from DNA sequences that the phylogeny has three branches for the three geographic groups of barbets and that the toucans are the sister group of the Neotropical barbet branch. Maintaining the principle of monophyly, we could recognize one family with four subfamilies (demoting toucans from family to subfamily rank) or designate all four groups as separate families. Most world lists have opted to elevate the three barbet clades to family rank, along with the family Ramphastidae.[15]

A final issue is the sequence of birds in lists and field guides. An accurate phylogeny of a group can justify more than one possible linear sequence. Trees 1, 2, and 3, shown here, carry exactly the same information about the relationships among species A, B, C, and D.

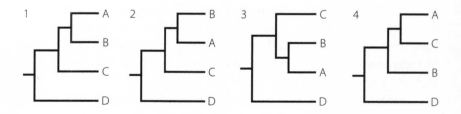

Tree 2 is the same as tree 1, with species A and B rotated at the branch point. But they are still each other's closest relatives, as they are also in tree 3. Tree 3 rotates C around its branch point with A + B, but it is still their closest relative. The three trees express the same information—but they are entirely different from tree 4, which describes species C rather than B as A's closest relative. If a linear list of species is to have any correspondence with their phylogeny, some sequences are equally good (such as A, B, C, D and B, A, C, D), while other sequences would violate the phylogeny (A, C, B, D would be misleading). So the makers of checklists have some flexibility in how radically to reorder species and yet keep closest relatives together. Until recently, several species of sandpiper—Greater Yellowlegs (*Tringa melanoleuca*), Lesser Yellowlegs (*T. flavipes*), and Willet (*T. semipalmata*)—were listed in that order, with the two very similar yellowlegs species together. The fifty-eighth supplement to the American Ornithologists' Society's checklist of North American birds reordered them, along with several Eurasian species that occasionally visit North America, in this sequence: (1) Lesser Yellowlegs, (2) Willet, (3) Spotted Redshank (*T. erythropus*), (4) Common Greenshank (*T. nebularia*), (5) Greater Yellowlegs, followed by (6) Common Redshank (*T. totanus*).[16] This was based on a molecular phylogenetic study by Rosemary Gibson and Allan Baker, at the Royal Ontario Museum in Toronto. Their phylogeny looks like this, with three branches forming a trichotomy that expresses the uncertain relationship between the Common Redshank and the other two groups of species.

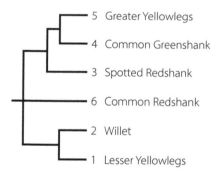

By rearranging the branches that form the trichotomy, we get a tree that shows exactly the same relationships among the species.

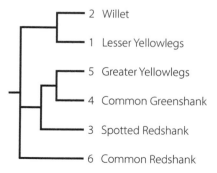

So, we could make a list that reflects the postulated relationships but keeps the two yellowlegs species happily together and brings smiles to the faces of all those birders who were so disappointed to see them divorced.

————

I started this chapter with changes in bird classification. This has led us into a range of interesting topics, such as how clarifying relationships among living birds can give us glimpses into the history of their evolution. We can learn something about the sequence in which different birds and their characteristics came into existence—without any fossils at all! But, of course, there is a fossil record that we can look to for information that we cannot find in living species. Using these two ways of peering into the past, let's look at a few aspects of the evolutionary history of birds.

PLATE 1. Small-billed (left) and large-billed (right) forms of the Black-bellied Seedeater (*Pyrenestes ostrinus*). (Courtesy of Thomas B. Smith.)

PLATE 2. An adult male Superb Fairywren (*Malurus cyaneus*). (Photo by Tobias Hayashi.)

Color versions of these images can be accessed at: https://press.princeton.edu/books /paperback/9780691264639/how-birds-evolve

PLATE 3. Campo Flicker (*Colaptes campestris*). (Courtesy of Dušan Brinkhuizen.)

PLATE 4. The South Island Takahe (*Porphyrio hochstetteri*), a large, flightless gallinule in New Zealand. A highly endangered species, it was once thought extinct. This banded bird is in a population introduced to a small island from a remote population that was discovered in 1948. (Photo by Douglas Gochfeld.)

PLATE 5. The Kakapo (*Strigops habroptila*), a highly endangered large, flightless, nocturnal parrot in New Zealand. (© Josep del Hoyo / Macaulay Library at the Cornell Lab of Ornithology.)

PLATE 6. Speckled Mousebird (*Colius striatus*). The six species of mousebirds are an old lineage now restricted to Africa. Their unique foot structure enables them to scamper along and hang from branches, as shown. (Photo by the author.)

PLATE 7. Lichtenstein's Sandgrouse (*Pterocles lichtensteinii*) in Eilat, Israel. Sandgrouse are related to pigeons. (Photo by Douglas Gochfeld.)

PLATE 8. Purple-crested Turaco (*Tauraco porphyreolophus*). Distantly related to cuckoos, the fruit-eating turacos today are restricted to Africa. (Photo by the author.)

PLATE 9. Black Skimmer (*Rynchops nigra*). The three species of skimmers, related to terns, have a unique bill, uniquely used. The lower mandible slices through the water as the bird flies and is snapped closed when it encounters a fish. (Photo by the author.)

PLATE 10. American Woodcock (*Scolopax minor*). (Courtesy of Phil Jeffrey.)

(A)

(B)

PLATE 11. (a) White and (b) brown morphs of the Red-footed Booby (*Sula sula*). (Courtesy of Daniel López-Velasco.)

PLATE 12. A Marbled Godwit (*Limosa fedoa*), with Semipalmated Sandpipers (*Calidris pusilla*) behind. The godwit's much longer bill, relative to body size, provides access to different food sources than the small sandpipers. (Courtesy of Steve Walter.)

PLATE 13. Red-faced and black-faced forms of the Gouldian Finch (*Erythrura gouldiae*). The bottom bird is a Long-tailed Finch. (Photograph by Mike Fidler Save the Gouldian Fund.)

(A)

(B)

PLATE 14. (a) Tan-striped and (b) white-striped morphs of the White-throated Sparrow (*Zonotrichia albicollis*), photographed in Central Park, New York City. (Courtesy of Michael D. Stubblefield.)

PLATE 15. Gray (left) and rufous (right) forms of the Tawny Owl (*Strix aluco*), held by a researcher. (Courtesy of Patrik Karell.)

(A)   **(B)**

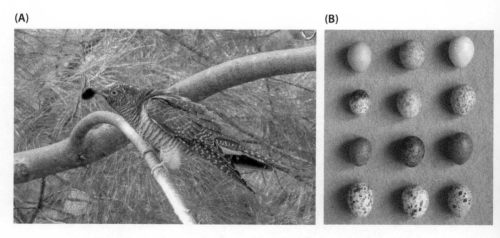

PLATE 16. Egg polymorphism in the Common Cuckoo (*Cuculus canorus*). (a) A Common Cuckoo. This individual, a remarkable vagrant, was discovered in Rhode Island, USA, in November 2020. (b) The left column shows eggs of four favorite host species (from the top): Dunnock (*Prunella modularis*), Reed Warbler (*Acrocephalus scirpaceus*), Meadow Pipit (*Anthus pratensis*), and Great Reed Warbler (*Acrocephalus arundinaceus*). The central column shows an egg of the cuckoo gens that specializes on each host. The match is quite close except for the Dunnock, which accepts any colored egg. The right column shows artificial eggs used by researchers to test rejection responses, painted to match the host species' eggs. ([a] courtesy of Lisa Nasta; [b] from Brooke and Davies 1988, with authors' permission.)

PLATE 17. The curious Hoatzin (*Opisthocomus hoazin*), the sole species in a lineage that diverged from other land birds more than 60 million years ago. (Photo by the author.)

PLATE 18. The Bar-headed Goose (*Anser indicus*) migrates at higher altitudes than perhaps any other bird. (Photo by Hannu Jännes.)

PLATE 19. The astonishing Sword-billed Hummingbird (*Ensifera ensifera*). (Courtesy of Dŭsan Brinkhizen.)

PLATE 20. In form and color, the Plumbeous Antbird (*Myrmelastes hyperythrus*) is a fairly typical antbird of the understory in tropical American forests. (Courtesy of Jon Irvine.)

PLATE 21. A Malleefowl (*Leipoa ocellata*) on his mound. (Courtesy of John Barkla, www.thewonderfulworldofbirds.com.)

PLATE 22. A researcher approaches a chick of the Northern Royal Albatross (*Diomedea epomophora sanfordi*) at the Royal Albatross Centre in Dunedin, New Zealand. Young birds are in the nest for ten months before they take their first flight. (Photo by the author.)

PLATE 23. An American Coot (*Fulica americana*) bearing food for its orange-plumed chick. (Courtesy of Bruce E. Lyon.)

PLATE 24. Siblicide in the Nazca Booby (*Sula granti*). The larger of two chicks is evicting its sibling from the nest. (Photo by Liz Leyden.)

PLATE 25. A pair of Greater Painted-Snipe (*Rostratula benghalensis*). The more highly colored female, in front, courts males and provides less care for the young than males do. (Courtesy of Markus Lilje.)

PLATE 26. A stunning male Wilson's Bird-of-Paradise (*Diphyllodes respublica*) on its display arena in West Papua. (Courtesy of Daniel López-Velasco.)

PLATE 27. Red Junglefowl (*Gallus gallus*) in Thailand. (Photo by the author.)

PLATE 28. One of the few sexually dimorphic species of parrot: female Eclectus Parrots (*Eclectus roratus*) are red and blue, while males are green. (Courtesy of Markus Lilje.)

PLATE 29. Four members of the genus *Tangara*: (a) Silver-throated (*T. icterocephala*), (b) Red-necked (*T. cyanocephala*), (c) Flame-faced (*T. parzudakii*), and (d) Plain-colored (*T. inornata*) Tanagers. The drab Plain-colored Tanager evolved from a more brightly patterned ancestor. (Photos courtesy of Dŭsan Brinkhuizen.)

PLATE 30. A "selfish herd": a flock of Black-tailed Godwits (*Limosa limosa*) in Thailand took flight when a Peregrine Falcon (*Falco peregrinus*) approached. (Photo by the author.)

PLATE 31. Two Wild Turkey (*Meleagris gallopavo*) brothers form a team to court females. (Courtesy of James Mott.)

PLATE 32. A White-throated Bee-eater (*Merops albicollis*), a species that beautifully supports the model of kinship-based altruism. (Courtesy of Daniel López-Velasco.)

PLATE 33. An Okinawa Rail (*Hypotaenidia okinawae*). (Courtesy of Bryan Shirley.)

(A)

(B)

PLATE 34. (a) Red Fox Sparrow (*Passerella iliaca*) in winter, Central Park, New York, with White-throated Sparrows. (b) Sooty Fox Sparrow (*P. unalaschensis*) in winter, Oregon. Formerly considered races of the Fox Sparrow, they are now widely considered separate species. Their breeding ranges, in the boreal forest zone of North America, do not overlap. ([a] Photo by the author; [b] courtesy of Richard C. Hoyer/Birdernaturalist.)

(A)

(B)

PLATE 35. Geographic variation in the color of male European Pied Flycatchers (*Ficedula hypoleuca*). (a) Males are black in Norway, far from the range of the Collared Flycatcher (*F. albicollis*). (b) Males are browner in the Czech Republic, where the species overlaps with the Collared Flycatcher. The evolution of a different color reduces the frequency of hybridization. (Courtesy of [a] Bjørn Aksel Bjerke and [b] Miroslav Král).

PLATE 36. A male Common Yellowthroat (*Geothlypis trichas*) in New York. (Courtesy of Phil Jeffrey.)

PLATE 37. A Shoebill (*Balaeniceps rex*) in a papyrus swamp in Uganda. A "stand and wait" predator on large fishes and other prey. (Photo by the author.)

PLATE 38. A mural on a school in Ghana features White-necked Rockfowl (*Picathartes gymno-cephalus*), at right. Tour groups that come to see this strange species are a significant source of the village's income. Plate 43 shows the other species of rockfowl. (Photo by the author.)

(A)

(B)

PLATE 39. (a) A Collared Trogon (*Trogon collaris*) in South America. (b) A Scarlet-rumped Trogon (*Harpactes duvaucelii*) in tropical Asia. ([a] courtesy of Dŭsan Brinkhuizen; [b] courtesy of Markus Lilje.)

PLATE 40. A Solitary Tinamou (*Tinamus solitarius*) in Brazil. (Photo by the author.)

PLATE 41. A Banded Broadbill (*Eurylaimus javanicus*) in tropical
Asia. (Courtesy of Markus Lilje.)

PLATE 42. In Papua New Guinea, a male Flame Bowerbird (*Sericulus ardens*) displays to a female who watches from within his bower—a display prop, not a nest, constructed by the male. David Attenborough's film of the extraordinary display may be found by searching "The Bowerbird's Grand Performance! Life Story/BBC." (Courtesy of Mark Kirkpatrick.)

PLATE 43. A Grey-necked Rockfowl (*Picathartes oreas*) in Cameroon. The two species of rockfowl (see plate 38) are an early branch of the group (Passerides) that includes familiar Northern Hemisphere families. (Courtesy of Markus Lilje.)

PLATE 44. A Hawaiian honeycreeper, the Iiwi (*Drepanis coccinea*). The long, curved bill is convergent with the bills of some other nectar feeders, such as certain hummingbirds and sunbirds, but these honeycreepers are adaptively divergent fringillid finches. (Photo by Douglas Gochfeld.)

PLATE 45. Spoon-billed Sandpiper (*Calidris pygmea*). This remarkable species, which nests in Russian tundra, is Critically Endangered because of hunting on the wintering ground and because coastal development in Asia has reduced habitat for migrants to stop and feed. (Courtesy of David Erterius.)

PLATE 46. The extinct Alagoas Foliage-gleaner (*Philydor novaesi*), which persisted until 2008 in small forest "islands" in a sea of sugar cane in northeastern Brazil. This photo was taken in 2001. (Courtesy of Ciro Albano, Brazil Birding Experts.)

PLATE 47. Lear's Macaw (*Anodorhynchus leari*). Although restricted to two main breeding colonies in northeastern Brazil, the population's status has improved, thanks to strong conservation measures. However, its main food, the fruit of a certain palm, is being reduced by clearing for agriculture. (Courtesy of Ciro Albano, Brazil Birding Experts.)

PLATE 48. Crested Ibis (*Nipponia nippon*). Formerly widespread in northeastern Asia, this species was reduced to a few individuals in China. The population has been increased by captive breeding and release in China and Japan. (Courtesy of Shailesh C. Pinto.)

3

After *Archaeopteryx*

HIGHLIGHTS OF BIRD HISTORY

More than one of my birding friends have asked me, "Is it true that birds descended from dinosaurs?" I answer that yes, I think the evidence is strong and convincing. (Whether or not we can ever be certain that something is absolutely "true" is a philosophical question that I'm not the one to answer.) I must admit that it's hard to think of a warbler or hummingbird as a dinosaur—although most ornithologists today will declare that "birds *are* dinosaurs."[1] That might be one reason we find the relationship intriguing. But more generally, the question shows our interest in history. We want to know about our ancestry, and perhaps the history of our town, our country, our culture. We know that to really understand anything in human affairs, such as the culture, religions, or politics of a country, we need to know history.

That's equally true of anything we observe about organisms, whether we look at the genetics of bacteria or the workings of the human brain. All have been shaped by their long history of evolution. Like any history, the course of bird evolution is fascinating and prompts countless questions. When did birds evolve? How did they get their distinctive features, such as wings? When and how did they become so wondrously diverse?

Our knowledge of history, whether of human languages and societies or of bird diversity, is very limited: it has to be reconstructed from physical fragments (manuscripts, ruins, fossils) or by comparing today's languages, cultural features, or species in order to infer a probable past. We glimpse the history of evolution from the fossil record and from phylogenetic comparisons of species.

We know that the fossil record is very incomplete (or at least very incompletely sampled) because paleontologists continue to find new forms of life. For example, Daniel Field and colleagues[2] reported in March 2020 the discovery of a 68-million-year-old fossil that appears to be related to the lineage that gave rise to the waterfowl and land fowl such as chickens. It is one of very few fossils showing that the clade of modern birds had already evolved by that time.

A paleontologist who studies birds knows the skeletal features of diverse birds at a level of detail that most people—including most biologists—would find astonishing, and uses this knowledge to compare fossil specimens with skeletons of both living and extinct birds in museum collections. For instance, Daniel Ksepka and collaborators studied a fossil that they assigned to the lineage that gave rise to the order Apodiformes—swifts and hummingbirds—because of its "abbreviated humerus," "blunt olecranon," "an ossified arcus extensorius of the tarsometatarsus," and other such features.[3] Because the fossil record of birds (and most other organisms) is so incomplete, we can seldom trace step-by-step the changes that culminate in new higher taxa (such as families or orders). The Ksepka team described this fossil as a member of the "*stem group*" of the order Apodiformes. A stem group, in contrast to the "*crown group*" with the derived features of living species, consists of extinct members of the clade with characteristics that are steps toward the modern form. Ksepka's fossil had longer legs than living Apodiformes, and it was unusual because its plumage was preserved, so the researchers could show that its wing shape was intermediate between swifts and hummingbirds. (Incidentally, Apodiformes, from the swift genus *Apus*, means footless or legless, in reference to these birds' extremely short legs.)

We can often use a "molecular clock" to estimate when certain events happened, such as how long ago two lineages split from a common ancestor or when a certain characteristic evolved. In many genes, the substitution of new for old base pairs has happened at a roughly constant rate, averaged over long periods of time.[4] So if the DNA sequence difference is 1% between species A and B, but 2% between species A and C, we might conclude that lineage C split off about twice as long ago as A and B separated from their common ancestor. But how long ago was that, in years? We need to know how fast the clock is ticking.

Rob Fleischer and colleagues estimated how fast some genes evolve in the Hawaiian honeycreepers, a spectacular radiation of birds that have evolved very different bills as adaptations to different ways of feeding.[5] They took advantage of the extraordinary geology of the Hawaiian Islands, which formed sequentially as the Pacific plate has moved northwestward over a "hot spot" of magma extrusion from the Earth's crust. The oldest island, Kauai in the northwest, is 5.1 million years old; the youngest, the "big island" Hawaii, at the southeast end of the chain, is 0.43 million years old. Fleischer and colleagues found that the basic framework of the honeycreepers' phylogenetic tree matches the island sequence. The oldest split in the phylogeny separates a species on Kauai from all the others, and the species restricted to the big island are most closely related to those on the nearest (and next oldest) island. Fleischer and colleagues related the percent difference in the sequence of a mitochondrial gene between a pair of bird species to the maximum possible age of the younger species (that is, the age of its island). They found that older pairs of species are more divergent than younger pairs and that the rate at which this DNA sequence diverges between species is about 1.6% per million years (figure 3.1). (So a single species undergoes about 0.8% change of its DNA sequence per million years.) In a later study, they estimated that the complete mitochondrial genome diverges at a rate of 1.8% per million years. They also studied some chromosomal genes, which evolve much more slowly.

A more common way of "calibrating" the rate at which the molecular clock ticks is to use fossils. A bird family is at least as old as its oldest known fossil, and this age is therefore the minimal one at which the family split from its closest relatives. An example is the oldest fossil frigatebird, carefully dated at 51.81 *Ma* (million years ago). This is the best estimate we have of when the frigatebird lineage separated from its closest relatives (a clade that includes gannets, boobies, and cormorants). This date, together with the percent DNA sequence divergence between these clades, provides a rate of sequence divergence per million years. It is one of the most reliable fossil calibrations for the rate of DNA sequence evolution in birds.[6] Most large-scale studies of bird phylogeny use several careful fossil calibrations of this kind to come up with an average rate of DNA sequence evolution, which researchers assume holds good for a major group of birds, or for birds in general.[7]

Pause and consider: we are talking about an astonishing advance in the progress and integration of science! I suspect few of my teachers in the 1960s imagined that we would be studying birds by combining information from geology and molecular biology—disciplines that are miles apart. There is beauty in the synthesis of science, as in synthesis of knowledge, whatever its nature.

Now that we have an idea of how we can glimpse evolutionary history, what about the history itself? We will return to DNA-based phylogenies, but the origin and early evolution of birds is the province of paleontologists.[8] What about dinosaurs?

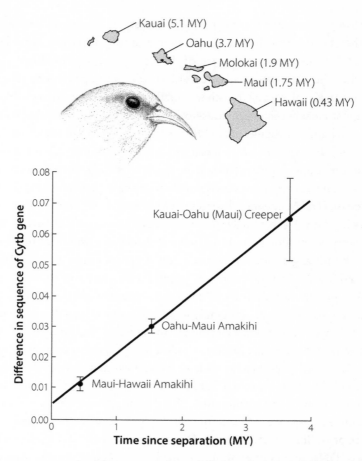

FIGURE 3.1. The difference in the DNA sequence of a mitochondrial gene between pairs of honeycreeper populations on different Hawaiian Islands, in relation to the age of the younger island in each pair. The islands Hawaii, Maui, and Oahu are 0.43, 1.75, and 3.7 million years (MY) old, respectively, so these are the maximal ages of their bird populations. The DNA sequence divergence between these three populations and their older, closest relatives is almost perfectly correlated with the populations' ages. An Amakihi (*Chlorodrepanis* species) is pictured. (After Fleischer et al. 1998; art, Luci Betti-Nash.)

Dinosaurs aren't just big, fearsome reptiles; they are a particular clade (branch) of archosaur reptiles that arose in the early Triassic period, about 240 Ma, and which also include extinct and living crocodiles. (So, crocodiles are the closest living relatives of birds.) And not all dinosaurs were big (or very fearsome): some were smaller than chickens. The salient feature of dinosaurs—one profoundly important for the origin of birds—is that unlike almost all other vertebrates, they were bipedal, walking on hind legs held vertically below the body instead of sprawling to the side, as in other reptiles.[9] Both this posture and the structure of the hind legs are unusual and are evidence that birds are modified dinosaurs.

As in humans, the leg bones include a basal femur (thigh bone) that joins the tibia and fibula (shank bones) at the forward-pointing knee; a set of upper and lower tarsal (ankle) bones; the metatarsals (long bones in our foot); and, finally, the digits (figure 3.2). The ancestors of all living tetrapod (four-legged) vertebrates had five fingers and five toes; many descendants have evolved lower numbers.

Like dogs and many other mammals, dinosaurs and birds walk on their toes, with the metatarsals elevated and the ankle forming a backward-directed joint. (A bird's ankle is usually visible, but its knee is usually covered by belly feathers.) Birds evolved from theropod dinosaurs, the group that includes the colossus *Tyrannosaurus rex*. A remarkable feature of many theropods is that the upper ankle bones are fused to the lower end of the tibia and fibula, just as in birds. If you eat a chicken drumstick, you will see a slender fibula and a more massive tibia—but this bone, technically called the tibiotarsus, actually includes ankle bones. Also in many theropods, the lower ankle bones became fused to the metatarsals and all of these became partly fused together. Later, they became fully fused in birds, forming a single tarsometatarsus bone. In both theropods and birds, toe 1 (our big toe) is directed backward (a condition that later enabled many birds to perch on narrow branches), and toe 5 was lost, so that they have three forward-directed toes.

The features that distinguish birds include more than leg structure. They have only three fingers, which lack claws, and have extensive fusion of wrist and digit bones. The ribs connect to the breast bone (sternum), which (in most birds) has a large keel on which wing muscles are anchored.[10] Many of the back vertebrae are fused with each other and with the pelvic girdle. Most of the vertebrae of the short tail skeleton are fused into a single bone, the pygostyle, that carries a fan of tail feathers. Modern birds lack teeth and have a horny sheath around the enlarged premaxilla, the movable upper beak.

The first evidence that birds evolved from dinosaurs came to light in 1863, when *Archaeopteryx lithographica*, from Jurassic deposits about 150 Ma, was discovered: a beautifully preserved specimen with dramatic impressions of wing feathers that are similar in form and organization to those of modern birds. Thomas Henry Huxley, known as "Darwin's bulldog" for his strong

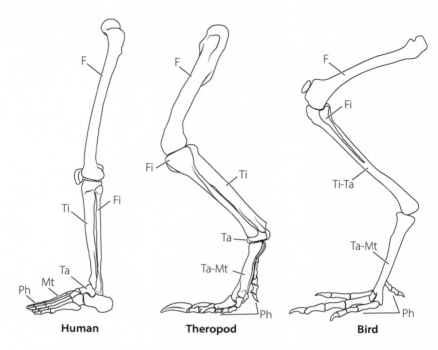

Human   **Theropod**   **Bird**

FIGURE 3.2. A comparison of the leg skeleton of a human, a theropod dinosaur, and a bird. F = femur, Ti = tibia, Fi = fibula, Ta = tarsal, Mt = metatarsal, Ph = phalanges. The separate ankle bones of humans and most other tetrapod vertebrates have been fused, in birds and theropods, with the lower end of the tibia and the upper end of the metatarsals. The metatarsals of birds are fused into a single bone. Birds lack separate tarsal bones (Ta) but have instead a tibiotarsus (Ti-Ta) and a tarsometatarsus (Ta-Mt). (Art, Luci Betti-Nash, after various sources.)

support of Darwin, recognized that the skeleton is similar (nearly identical, in fact) to small theropod dinosaurs. Unlike modern birds, *Archaeopteryx* had teeth, claws on the long fingers, and a long tail skeleton with feathers on either side (figure 3.3A). It lacked some of the important features of modern birds, such as the large breast bone (sternum), fusion of the ribs to the sternum, fusion of vertebrae, and the horny beak sheath. In the 1970s, John Ostrom, a paleontologist at Yale University, described a new theropod dinosaur, *Deinonychus* (a close relative of *Velociraptor*, a turkey-sized predator popularized by the film *Jurassic Park*).[11] Ostrom noted the features it shared with *Archaeopteryx*, such as the form and orientation of the fingers and the fusion of two wrist bones into a unique form that enables a bird to fold its hand back along its forearm[12] (figure 3.4).

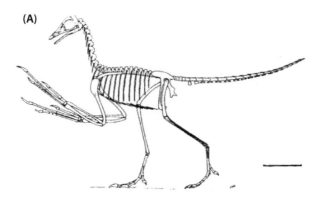

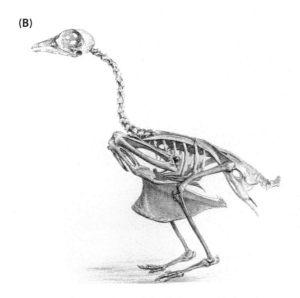

FIGURE 3.3. A comparison of the skeletons of (A) *Archaeopteryx* and (B) a modern bird, Pallas's Sandgrouse (*Syrrhaptes paradoxus*). Compared with *Archaeopteryx*, the modern bird lacks teeth and has an expanded cranium; fusion of the second and third fingers and fewer digit segments; loss of finger claws; a greatly expanded breastbone (sternum); ribs with overlapping processes; extensive fusion of the back vertebrae and fusion of these with the pelvis; and a shorter tail, with the terminal vertebrae fused into a large bone (pygostyle) that supports the tail feathers and the muscles that control them. (A from Ostrom 1976, by permission of the Linnean Society; B from Parker 1862.)

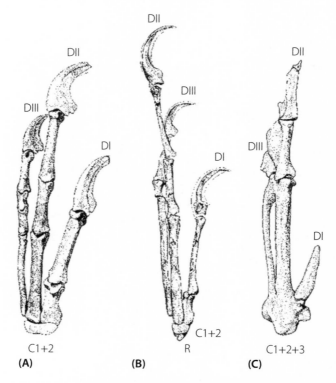

FIGURE 3.4. A comparison of the hand skeleton of (A) the theropod *Deinonychus*, (B) *Archaeopteryx*, and (C) a modern bird (a tinamou, *Nothura maculosa*). *Deinonychus* and *Archaeopteryx* have only three digits and share a large wrist bone (C 1 + 2) that enabled them to flex the hand backward. These were among the points of evidence that birds evolved from theropod dinosaurs. The modern bird's hand lacks claws, and the digits are highly modified by reduction and fusion of elements. (From Wagner and Gauthier 1999, copyright [1999] National Academy of Sciences, USA.)

Since Ostrom's work, many other bird relatives have been described, especially in China. We now know that birds and extinct theropods share many features, some still being discovered. For example, Jasmina Wiemann and her colleagues have chemically analyzed the egg shells of several groups of dinosaurs and reported in 2018 that red-brown and blue-green pigments, which occur in the eggs of birds only, evolved uniquely in theropods.[13] Birds and fossil theropods have the same leg structure and hand structure; both have hollow airspaces in some of their bones, and—best of all—theropods and many other groups of dinosaurs had feathers: unique, complex structures that surely have evolved only once in the history of life.[14] At first, feathers were simple in form and may have served as insulation to maintain body temperature. (Whether or not dinosaurs were "warm-blooded" is still debated.[15]) Some theropods had more advanced feathers, with a vane made up of interlocking barbs, and some (members of the Paraves in figure 3.5) almost certainly flew.

The various features of birds evolved sequentially over the course of about 150 million years. In figure 3.5, we see the successive evolution of (1) bipedal locomotion and (2) simple filamentous feathers in Dinosauria; (3) clavicles fused into a wishbone in Theropoda; (4) vaned feathers, adapted to flight, and the ability to fold the hand backward in a clade named Maniraptora; (5) wings and advanced flight in Paraves; (6) fusion of several tail vertebrae into a pygostyle in Pygostylia; and (7) a large, keeled breastbone, as well as a kinetic skull and rapid growth, in Ornithuromorpha: the Cretaceous and modern birds. This is a good example of "mosaic evolution," meaning that different characteristics evolve at different rates, not in lock step.[16] It shows that there is no firm distinction or divide between "nonavian" dinosaurs and birds.

After the basic "body plan" of birds evolved, there was a burst of rapid evolution.[17] During the Cretaceous period (145–66 Ma), birds proliferated into many forms (such as the "opposite birds," Enantiornithes, named for the structure of the wing socket) that became extinct by the middle of the period.[18] A few water birds are known from the late Cretaceous, such as *Ichthyornis* and the flightless *Hesperornis*, which show some important steps toward modern birds. Specifically, modern birds have a unique hinge between the skull and the upper jaw (maxilla and premaxilla bones) and use movable palate bones to open the bill by raising the upper jaw. *Ichthyornis* shows an intermediate step toward this "kinetic" skull.[19] Both *Ichthyornis* and *Hesperornis* still had teeth. In living birds, tooth primordia start to grow in very young embryos and then stop. A mutation is known in chickens that causes somewhat more advanced tooth rudiments to form, and this also happens when chick jaw cells are influenced by implanted jaw tissue from a mouse embryo. But although most of the genes needed to develop dentine and enamel are still present in a bird's genome, they are pseudogenes: "fossil genes" that have accumulated many mutations that make them nonfunctional.[20]

Many of the differences between modern birds and their theropod ancestors are adaptations for flight. The hand, bearing the wing's primary feathers, has a drastically altered structure: most of the wrist and metacarpal bones are

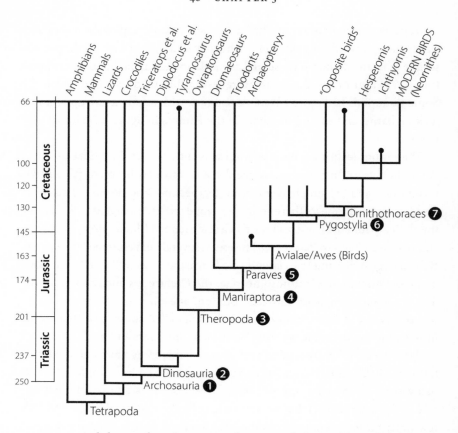

FIGURE 3.5. A phylogeny of vertebrate groups, focusing on the lineages that include common ancestors in the history of birds. Among the names of the successively more derived clades, notice the Avialae (Aves) in the late Jurassic, whose descendants include the modern birds in the Ornithurae. The numbers below the tree mark the sequential evolution of (1) bipedal posture and (2) feathers in Dinosauria, (3) wishbone in Theropoda, (4) vaned feathers in Maniraptora, (5) wings in Paraves, (6) fused tail bones (pygostyle), and (7) keeled sternum in Ornithothoraces. (After Brusatte et al. 2015.)

fused together, and there are only three digits, each reduced to only one or two segments. Many parts of the skeleton are fused to resist the stresses of flight: the pectoral girdle has the clavicles fused into a wishbone, and the rear back vertebrae and the anterior tail vertebrae are fused into a single large bone that is fused to the pelvis. The short tail ends with several vertebrae fused to form the pygostyle, supporting tail feathers that are spread and lowered to brake the bird's flight when it alights. Each rib has a projection that overlaps the following rib, strengthening the rib cage. Perhaps to reduce weight, female birds have only one ovary, and the male's testes are very small except during the breeding season, when they enlarge greatly.

And now we come to modern birds, to the end of the Cretaceous period, to one of the most dramatic events in the history of life, and to the conjunction of the fossil record and molecular phylogeny in piecing together birds' history.

The end of the Cretaceous period and the start of the Cenozoic era, 66 Ma, are marked by the second most severe extinction event in history. As many as 75% of the species on Earth were extinguished, mostly owing to the impact of an asteroid or comet that created a pall of dust that darkened the Earth and reduced temperature and photosynthesis. This is called the K–T extinction or the K–Pg extinction. ("K" is for Cretaceous; "T" is for Tertiary, the first period of the Cenozoic in an older version of the geological record; "Pg" is Paleogene, the first period in the current version.) No nonavian dinosaurs survived, nor did most of the Cretaceous lineages of avian dinosaurs (birds). Like many other groups, birds suffered mass extinction: all living birds are descended from the very few lineages that survived.

There is strong molecular and anatomical evidence that modern birds (Neornithes) form two main clades, the Palaeognathae (ostriches, tinamous, and relatives) and Neognathae, which, in turn, are divided between Galloanseres (screamers, ducks and geese, and galliform birds such as pheasants and quail) and all others (Neoaves). The Neornithes were known from the Cretaceous fossil record by only one species, a duck-like galloanserine, until the 2020 discovery of the 68-million-year-old fossil that I mentioned earlier. The almost complete lack of Neornithes fossils suggests that this group originated late in the Cretaceous and that very few Cretaceous species survived the extinction event. There is reason to think that these few were groundliving and aquatic forms and that Cretaceous tree-living birds were extinguished by the reduction of forests.[21]

After the K–Pg extinction event, the earliest bird fossils, near the K–Pg boundary, belong to "stem groups" that later gave rise to modern loons, penguins, and other water birds. Among land birds, one of the first, dated at 3.5 million years after the mass extinction, is related to the small, arboreal, fruiteating African mousebirds (plate 6). DNA-based phylogenies of living birds (figure 3.6) show that mousebirds branched off from a large clade of diverse land birds (kingfishers, hornbills, woodpeckers, trogons, and others) that is the sister group of another, even more diverse clade of land birds (falcons, parrots, and passeriforms). This suggests that within a few million years after the K–Pg extinction there were diverse lineages of land birds, the ancestors of many of our familiar living orders. It appears that in birds, just as in mammals, fishes, and many other groups,[22] there was an explosive adaptive radiation, or diversification. Santiago Claramunt and Joel Cracraft and research groups led by Erich Jarvis and Richard Prum all found that the amount of DNA sequence divergence between the sequential origin of different orders of birds is very small, showing that they split from their common ancestors within a short time (see figure 3.6).[23] Splits leading to many of today's orders of birds—highly diverse in size, structure, habitat, feeding habits, and breeding biology—happened within the 10-million-year-long Paleocene, the first geological epoch of the Cenozoic era that followed the K–Pg catastrophe.

The studies led by Jarvis and Prum are the most comprehensive studies so far. Because it can be extremely hard to determine relationships among lineages that were sequentially formed within a very short time, the phylogenies produced by these two studies have some important differences, although they agree on many of the relationships. Here are some interesting highlights in the Prum et al. phylogenetic tree (see figure 3.6).

First, Prum and Jarvis agree with previous authors that the common ancestor of all living birds gave rise to two species that were the ancestors of the palaeognaths (ostriches and others) on one hand, and of all other birds (neognaths) on the other hand. The next oldest split, as I already mentioned, is between the Galloanseres (land fowl and waterfowl) and the Neoaves (everything else). It's interesting that the groups that split off first are ground- or water-living forms, suggesting that this was the condition of the ancestors of all living birds. That condition is consistent with the decimation of forests by the asteroid impact, which may have "selected against any flying dinosaurs . . . committed to arboreal ecologies."[24]

Within the Neoaves, the first branch in the Prum phylogeny includes the nightjars and relatives, the swifts, and the hummingbirds. These are all aerial feeders that don't use their feet for much except perching; there was earlier evidence that they are related to each other, but no one had suggested they might be an early, separate branch from all other Neoaves (nor did Jarvis et al.). Next, Prum et al. propose a group that includes two pairs of families that have long been suspected to be relatives: pigeons + sandgrouse and cuckoos + turacos (plates 7, 8).

A big branch, on which Prum and Jarvis mostly agree, is remarkable: it consists of an astonishingly diverse set of water birds—*all* the water birds except the ducks, dippers, and a few rails and their relatives. Shorebirds, gulls and

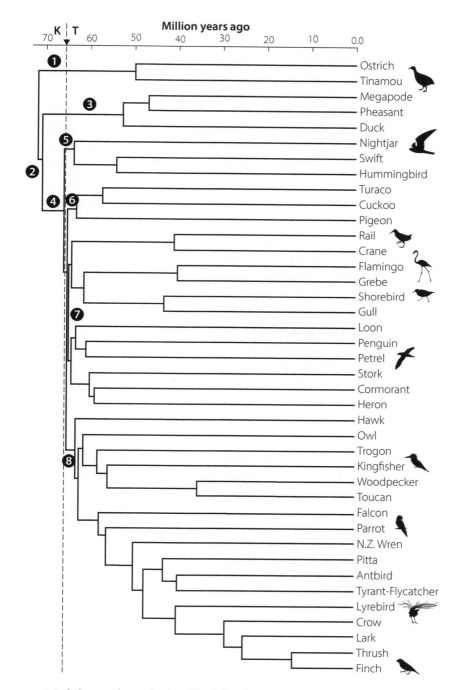

FIGURE 3.6. A phylogeny of some families of birds, based on DNA sequences at a genomic level. Some of the major clades mentioned in the text are indicated by circled numbers (1, Palaeognathae; 2, Neognathae; 3, Gal-loanseres; 4, Neoaves; 5, Strisores; 6, Columbaves; 7, Aequorlitornithes; 8, Inopinaves). Families are denoted by the English names of well-known representatives. The vertical line at 66 million years ago is the boundary between the Cretaceous (K) and the Paleogene (Pg). I refer to this phylogeny in later chapters of this book. (After Prum et al. 2015; silhouettes, Luci Betti-Nash.)

terns, loons, tubenoses, penguins, herons, pelicans, cormorants, storks, grebes, flamingos—they are all here. They presumably all trace their forebears to a water-living ancestor in the early Paleocene, when many of the splits occurred that gave rise to these modern groups. Within these water-loving "Aequorlitornithes" or "Aequornithia" (newly coined, not very lyrical, Latin-Greek hybrid names[25]), some relationships have been known or suspected for a long time—penguins are related to petrels and other tubenoses, for instance. Some relationships are less obvious: flamingos and grebes are closest relatives. A few members of this huge assemblage, such as the pratincoles and some plovers, have become denizens of dry habitats such as grasslands, but most of the others have retained a watery lifestyle for more than 60 million years.

The ancestral lineage of the "Aequorlitornithes" diverged, soon after the K–Pg extinction, from the ancestral lineage of an enormous assemblage of land birds that Prum et al. call the "Inopinaves": the "unexpected birds." Well named—some of the relationships (such as the position of the hawks as a sister group of almost all the others) had never been suggested before. One great group (Coraciimorphae) includes five traditional orders: owls, mousebirds, trogons, Coraciiformes (rollers, hornbills, kingfishers and others), and Piciformes (puffbirds, barbets, toucans, woodpeckers). The other great group in the Inopinaves has four clades: the seriemas (two South American species of long-legged, walking predators); then the Falconiformes (falcons and caracaras, not closely related to the hawks they so resemble); then the parrots and their sister group, the order Passeriformes: the perching birds that constitute half of the living species of birds. These relationships were a surprise when they were first detected in DNA studies by research teams headed by Per Ericson in 2006 and by Sharon Hackett in 2008, but they have held up in all studies since then.[26] As the Prum team remarks, the relationship of predatory seriemas and falcons to parrots plus passerines, of owls to the coraciimorphs, and of hawks to the rest of the land birds suggests that the entire Inopinaves assemblage arose from predatory ancestors that repeatedly gave rise to lineages with other feeding habits.

The order Passeriformes (perching birds, or "passerines"), with more than 5,000 species, accounts for half the world's birds. How did they become so diverse?

The DNA evidence confirms the traditional division of most "perching birds" into two suborders (see figure 2.3), the suboscines (Tyranni) and the oscines (Passeri).[27] A few suboscines—pittas and broadbills—occur in the Old World tropics (Africa, Asia, and Australia), but tropical America harbors most of them: a profusion of tyrant flycatchers, antbirds, ovenbirds, cotingas, manakins, and others. There are about four times as many species of oscines as suboscines. They have much more complex syrinx ("voice box") muscles and tend to have more complex songs than suboscines.[28] It's possible that the greater complexity of oscines' songs has led to their higher diversity.

But plenty of other factors contribute to the huge diversity of oscines, and of the Passeriformes as a whole. Many passerine groups have undergone spectacular *adaptive radiation*, a term used to describe the evolution of ecological diversity within a rapidly multiplying clade.[29] They have radiated into diverse ecological niches, becoming different in feeding habits and the habitats they occupy. Most passerines eat insects, fruit, seeds, nectar, or some combination, and each of these habits has evolved multiple times. Table 3.1 gives examples of each type, among the suboscines and among oscines, in three regions of the world. The examples in table 3.2 show that methods of foraging, especially for insects, have also evolved independently many times.

Birds have long provided some of the best examples of how "ecological opportunity" encourages adaptive radiation. Species have an opportunity to diversify in food and habitat use if these resources aren't already pre-empted by other species. Some of the best examples are on islands, and the most famous example, by far, is the group of birds in the Galápagos Islands that the British ecologist David Lack named "Darwin's finches" (see figure 1.4). The size and shape of the bill differs among the insectivorous warbler finches, the vegetarian finch (which eats fruits and buds), tree finches that excavate arthropods from dead wood, the tool-using woodpecker finch that uses a spine or twig to extricate arthropods, the cactus finch, and six species of seed-eating ground finches with bills that vary greatly in size and stoutness. One of the ground finches, known as the Vampire Ground Finch (*Geospiza septentrionalis*), supplements its diet with the blood of boobies![30] In the virtual absence of competition with other species, this group of finches has evolved in the last 4 million years to exploit almost every possible food source.

Darwin's finches are a microcosm of the adaptive radiation of birds as a whole. Much of the immense diversity of bird bills originated very early in the evolution of Neoaves, as we see in the characteristic differences among orders and families:[31] dagger-like bills of herons; short, hooked bills of raptors; slender, sediment-probing bills of sandpipers; modified bills of flamingos and shovelers that sieve plankton from water; serrated bills of mergansers and hook-tipped bills of cormorants, both adapted to grasp fish.

TABLE 3.1. Parallel Diversification of Passerines' Diet

	Diet			
	Insects	Fruit	Seeds	Nectar
Suboscines	Tyrant flycatchers (Tyrannidae)	Manakins (Pipridae)	Tapaculos (Rhinocryptidae; not specialized)	—
Oscines: New World	Wood warblers (Parulidae)	Tanagers (Thraupidae, part)	Sparrows (Passerellidae)	Honeycreepers (Thraupidae, part)
Oscines: Africa, Eurasia	Warblers (Sylviidae and others)	Bulbuls (Pycnonotidae)	Finches (Fringillidae), Waxbills (Estrildidae)	Sunbirds (Nectariniidae)
Oscines: Australia and New Guinea	Thornbills (Acanthizidae)	Birds-of-paradise (Paradisaeidae)	(Old World estrildids)	Honeyeaters (Meliphagidae)

TABLE 3.2. Parallel Diversification of Passerines' Modes of Foraging

	Foraging mode				
	Aerial (flycatching)	Foliage of trees	Tree trunks	Brush, scrub	Ground
Suboscines	Tyrant flycatchers (Tyrannidae, part)	Foliage-gleaners (Furnariidae, part)	Woodcreepers (Dendrocolaptinae)	Antbirds (Thamnophilidae)	Antthrushes (Formicariidae)
Oscines: New World		Wood warblers (Parulidae)	Nuthatches (Sittidae, shared with OW[a])	Wrens (Troglodytidae)	(Larks, pipits, from OW[a])
Oscines: Africa, Eurasia	Flycatchers (Muscicapidae, part)	Tits (Paridae)	Creepers (Certhiidae)	Babblers (Timaliidae)	Larks (Alaudidae)
Oscines: Australia and New Guinea	Wood-swallows (Artamidae)	Acanthizidae (Gerygones)	Australian creepers (Climacteridae)	Atrichornithidae (scrubbirds)	Quail-thrushes (Cinclosomatidae, part)

[a] OW = Old World.

Birds have also diversified by spreading into new regions and adapting to different climates. Using published data on the geographic range of 7,657 species of birds throughout the world, Christopher Cooney and coauthors found that families and genera that live in a wider range of climatic environments have more species. They estimated the rate at which each family or genus has diversified: the number of species divided by the time since their common ancestor (using a molecular clock with the phylogeny). The diversification rate was higher in the climatically more widespread groups.[32]

As we develop a more complete phylogeny of birds, we learn more about the evolution of their characteristics. A major theme is convergent evolution: similar evolutionary changes that evolve independently in different lineages. For example, loons, grebes, penguins, and cormorants (together with anhingas) have independently evolved to dive and pursue aquatic prey. In all of these lineages, the bones are less pneumatic—they have reduced air spaces—than in their relatives, which retain the pneumatic condition of their ancestors.[33] Denser bone makes these birds less buoyant, so they need expend less energy to stay under water. This is an example of the "phylogenetic comparative method," one of the ways in which biologists can test a hypothesis about whether and how a certain characteristic is an adaptation. There are many familiar examples. Swifts, swallows, and falcons have independently evolved long, slender, pointed wings that for aerodynamic reasons (such as shedding turbulent air eddies) equip the birds for fast flight. Species of pipits, buntings, sparrows, larks, and quail that live in grasslands all have streaky patterns that match their background and make them hard to see. Hawks, owls, and falcons have independently evolved the hooked beak, strong feet, and sharp, curved claws that we associate with raptors. These correlations tell us the adaptive raison d'être of these features. Sometimes the theme of convergent evolution remains but with variations. Among aquatic birds, webbed feet have evolved at least four or five times, but grebes, coots, and sungrebes have all independently evolved lobed toes instead. Nectar feeding has evolved many times and is associated with a long, slender bill in many hummingbirds, sunbirds, honeyeaters, tropical American honeycreepers (modified tanagers), and Hawaiian honeycreepers (modified finches). But the bill isn't modified in some parrots (lorikeets) that feed on nectar. They do have a brushy tongue that functions like the honey dipper you might have used to drip honey onto toast.

Convergent evolution provides some evidence bearing on a big question about evolution: how predictable is it? If some intelligent aliens had watched the first four-legged vertebrates colonize land about 375 Ma, would they have predicted the evolution of birds? Of humans?

George Gaylord Simpson, an eminent paleontologist at Harvard University, grappled with this question in 1964, during the early years of SETI: Search for Extraterrestrial Intelligence. He concluded that the evolution of humans had required so many improbable events that the existence of "humanoids" was extremely improbable (much less the chance of communicating with any that might exist). In a recent book on this theme, evolutionary biologist Jonathan Losos, likewise, asks if a reflective platypus might expect platypusoids to inhabit other plants, and answers, "Alas, no. Were we destined to be here? Hardly."[34]

I don't know how similar to birds a creature would have to be for us to call it an "avioid" or an "ornithoid." If we were satisfied with a flying vertebrate, the answer is that that was so likely to happen that it already has—twice (pterosaurs and bats). But if we want an avioid to be bipedal with feathers, toothless kinetic jaws, highly developed vision, a gizzard, and a high constant body temperature, I'm with Simpson and Losos: I think it is very unlikely indeed. A few lizards can briefly run on their hind legs, and a few mammals progress by bipedal hopping (kangaroos, jerboas, a few other rodents), but only humans and a few archosaur lineages—especially the dinosaurs—evolved habitual bipedal walking. And among the hundreds of thousands of species of terrestrial vertebrates that have existed since the Devonian, no other animal has evolved anything like the basic structure of feathers.

But birds' many forms and lifestyles are another matter; many of them have evolved multiple times and seem to argue against Simpson's view. Maybe they were inevitable, once the bird "body plan" had evolved. Large, flightless birds include familiar ratites such as ostriches but also the extinct *Diatryma* (related to geese and ducks) and the extinct Phorusrhacidae, the "terror birds" that dominated South American plains for millions of years, eating small horses and other animals. (I will return to the flightless ratites, and this theme, in chapter 11.) Aerial insect eating evolved independently in swifts, nighthawks, pratincoles, and swallows, all with similarly shaped wings.

It seems that convergent evolution has been so common in birds that there aren't many constraints on what can evolve into what—even some shifts that seem unlikely. All the nightjars and their relatives feed on insects, except for the Oilbird (*Steatornis caripensis*), which eats fruit. So does the Palm-nut Vulture (*Gypohierax angolensis*), alone in the great order of eagles, hawks, and kites. Among about 6,000 species of Passeriformes, one small lineage—the dippers—became truly aquatic. And I mentioned in chapter 2 the peculiar Ground Tit, so unlike its arboreal relatives. So are Simpson and Losos (and I) wrong? Was every avian form and way of life inevitable?

Not quite, I think. There are some striking one-off events in bird evolution. Only the megapodes incubate their eggs by burying them (chapter 7). Only kiwis have nostrils at the end of the bill (and a way of life to match). Only the Hoatzin (*Opisthocomus hoazin*) uses bacteria to ferment a diet of leaves (chapter 6). Only pigeons and doves produce food for their young—"crop milk." And anyone who has seen a skimmer (plate 9) has seen a bird like no other![35]

Every species has a legacy of characteristics that were formed in long distant ancestors and which limit its possible paths of evolution. Biologists call these "phylogenetic constraints." Unlike other fish-eating birds, Ospreys (*Pandion haliaetus*) catch fish with their feet instead of their bill—because that's how their clade (Accipitriformes) has caught prey for at least 40 million years. Even though some of the water birds in the Aequorlitornithes nest in trees, none has evolved to hop about and find food in trees in their 65-million-year history. Among the enormous assemblage that Prum and collaborators named "Inopinaves," the only aquatic lineage, the dippers, has a long way to go before its descendants converge with ducks or loons. The structure of swifts and hummingbirds is so extremely modified for flight—they are hardly able to hop—that is hard to imagine that they could ever have substantially different descendants. And for birds as a whole? Do they have any limits?

There is one way of life that has evolved in various groups of salamanders, frogs, lizards, snakes, and mammals, but never in birds. Many birds use their bills and feet to dig nest burrows, but none lives underground like a mole or a mole salamander. I imagine quite a few characteristics might make that a hard transition, but the most obvious is that the wing skeleton is so highly adapted for flight that evolving it into a digging tool, like a mole's hand, would be really unlikely. And how likely is it that the first steps toward underground life would be advantageous, considering that most birds use vision to find food? So—hark!—I predict that no bird lineage will evolve a burrowing lifestyle. I admit that's not a very scientific prediction, considering that 10 or 20 million years from now no one will be checking it. But it's a safe one.[36]

4

Finches and Blackcaps

HOW BIRD POPULATIONS CHANGE AND ADAPT

Among my early experiences as a young birder was an American Woodcock (*Scolopax minor*) in a wooded sector of Bronx Park in New York. It flew up in front of me and landed in plain view a short distance away but was invisible in the leaf litter. I detected it only when I spotted its large black eye and then slowly made out its outline, amazed by how it matched its background. Not long afterward, I found a dead woodcock and brought it home. My father had a taxidermist mount the bird, and I have vivid memories of its intricate, delicate pattern: some feathers black with rufous edges, others pale blue gray or cinnamon (plate 10). I saw beauty and dimly recognized function—what I would soon learn is called adaptation.

One reaon that birds are so fascinating is the immense range of their truly marvelous adaptations. The cryptic color and pattern of a woodcock is an intricate way of eluding predators, as it is in nightjars, pipits, plovers, and countless other species. A Barn Owl (*Tyto alba*) locates prey inerrantly in total darkness; a male Emperor Penguin (*Aptenodytes forsteri*) stands on ice for two months, incubating an egg, during the dark and bitterly cold and windy Antarctic winter; a Bar-tailed Godwit (*Limosa lapponica*) migrates nonstop for 11,000 kilometers from Alaska to New Zealand. I have already mentioned the extraordinary bill and feeding behavior of skimmers.

In Darwin's time, the features that equip animals and plants for their environment and way of life were taken as evidence of the wisdom and benevolence—and existence—of the Creator. The Reverend William Paley, in 1802, famously likened organisms to a watch, which by the intricacy of its structure in service to a clear purpose indisputably (he thought) points to an intelligent artificer. The realization that the exquisite functional features of organisms could be explained by a purely material process, with no necessity for an intelligent, supernatural designer, was one of the most important events in the history of human thought. This was the process of natural selection that Charles Darwin conceived in 1838, that Alfred Russel Wallace independently conceived in 1858, and that Darwin introduced in *On the Origin of Species* in 1859. In replacing supernatural explanation by a purely natural explanation that could be studied and tested, Darwin and Wallace made biology a science. Indeed, they helped to shape the nature of science by separating it from theology. The philosophical implications were profound. For example, the theological Western tradition taught that plants and animals existed for purposes—some for their usefulness or value to humans, others because they filled what would otherwise be gaps in a "great chain of being," a perfect creation.[1] But Darwin's theory rejected any grand purpose and any forethought: science finds no evidence for any design, goal, or purpose in the living world. All of science now takes this position; chemists do not envision a purpose for chemical bonds, nor astronomers for supernovas.

What is natural selection? Darwin noted that many features vary from one individual to another within a species and that the variations are often inherited. He proposed that "in the course of thousands of generations," some useful variations would arise and that these individuals would have "the best chance of surviving and procreating their kind," whereas "any variation in the least degree injurious would be rigidly destroyed" by the death of affected individuals. "This preservation of favourable variations and the rejection of injurious variations, I call Natural Selection." Slight changes effected by natural selection, repeated and accumulated over vast spans of time, could result in such striking adaptations as the skimmer's bill or the avian feather.

So, evolution by natural selection occurs if (1) individuals in a population vary in a characteristic, or trait; (2) the trait is correlated between parents and their offspring (that is, the variation is inherited); and (3) there is a correlation between individuals' characteristics and their average survival or reproductive rate (that is, there exists natural selection). If these conditions hold, the proportion of individuals with an advantageous trait (say, darker color) may increase from generation to generation. Now that we (unlike Darwin) know about genes, we know that underlying the change in the trait (the *phenotype*) is a change in the proportion of individuals in the population that carry different versions (different *alleles*), which arose by mutation, of genes that affect the characteristic. If darker individuals have higher average survival than pale ones, they will produce more offspring, on average, and these offspring tend to inherit the alleles that produce dark color. So the frequency—the proportion—of "dark alleles" and of dark individuals increases. If this happens again and again in successive generations, the frequency of dark alleles gradually increases and reaches 100% (when the dark allele is said to be *fixed*). Exactly the same process occurs if darker and paler individuals differ not in survival but in their average reproductive rate (number of offspring).

The elementary process of natural selection is simplicity itself: *natural selection is nothing more than an average difference in the rate of reproduction between two or more variant types of organisms or genes.* But despite its simplicity, it is widely misunderstood.[2] Natural selection isn't a beneficent, caring force—it isn't "Mother Nature" helping her children through life's trials. Organisms don't perceive a need to adapt and then change themselves to fit. In fact, evolution doesn't mean that individual animals and plants change from an "old" to a "new" form during their lives; instead, some are born with variant genes that alter one or more characteristics. The variant types of genes (and of features that are affected by the genes) do not originate in response to need: they arise by random processes of DNA mutation, whether the circumstances would make them useful or not. The most famous example of observed evolution is the replacement of pale-gray peppered moths by a black variant during the Industrial Revolution in England. The mutation that makes the moth black had occurred countless times in the species' history, as we know from preindustrial museum specimens. But it had always been kept rare because it was disadvantageous: it made black moths more visible to insect-eating birds since they stood out on the pale, lichen-coated tree bark. The black mutation became advantageous when pollution darkened tree trunks: now it made the moths less visible to insect-eating birds, and so it increased greatly.[3] Advantage depends on context.

Biologists refer to survival plus reproduction as fitness. The fitness of a genotype is the average number of offspring that individuals with that genotype produce during their lifetime.[4] As a simple example, suppose that among many newborn individuals with the same genotype, the chance (probability) of surviving to reproduce is 40%, or 0.4, and a survivor has 10 offspring, on average. Then the fitness of that genotype is $0.4 \times 10 = 4$. Another genotype might have a lower survival rate yet higher fitness if survivors have more offspring. If survival is 0.2 and reproduction is 25, fitness is $0.2 \times 25 = 5$; this genotype would increase and replace the other, and we would refer to whatever feature enables it to be so fecund as an adaptation.

Neither Darwin nor Wallace described any real examples of natural se-lection because no one had made the necessary studies. We now have many examples of natural selection in action and of evolutionary change wrought by natural selection within one or a few human lifetimes. The key ingredients of evolution by natural selection are inherited variation in species' character-istics; a correlation between a characteristic and survival or reproduction; and finally, changes across generations. All have been described in various birds.

First, what about inherited variation? Just as in humans, birds provide ex-amples that fall roughly into two categories: discrete and continuous variation. (These are not absolutely separate; variation is sometimes quasicontinuous.) Discrete variation, in which most individuals can be assigned to separate cat-egories, is called *polymorphism* (from Greek: many forms).

Birders in many parts of the world are familiar with species that have differ-ent color morphs, or "phases." The Snow Goose (*Anser caerulescens*) has white and "blue" forms; Tawny Owls in Europe are red or gray; Rough-legged Hawks (*Buteo lagopus*) and Parasitic Jaegers (*Stercorarius parasiticus*) are either dark or pale (figure 4.1). We now know that in several of these and other species, the dark/pale difference is caused by mutations in the same gene that have happened independently: the *melanocortin-1-receptor*, or *MC1R*, gene. What's more, this gene is the basis of color variations in several species of liz-ards and mammals, including humans! (Mutations in *MC1R* underlie red hair in human populations.)

The DNA sequence of the dark and pale alleles has been determined for several species. Patricia Baião and her colleagues[5] studied this gene in the Red-footed Booby (*Sula sula*), which has three color morphs (plate 11). They found variable base pair sites that encode amino acids in positions 85 and 297. (Recall from chapter 2 that each codon, or triplet of three DNA base pairs, can encode 1 of 20 amino acids—the small molecules that form the protein chain and determine its chemical activities; see figure 2.2.) Using letters for the amino acids, position 85 can have V or M, and site 297 can have H or A. These are inherited together, so an individual DNA sequence has either V-H or M-A:

The white morph of the booby has V-H in both of its gene copies (it is *ho-mozygous*); the brown morph has M-A in both gene copies; and the white-tailed brown morph has one V-H copy and one M-A (it is *heterozygous*). The closely related Nazca Booby (*Sula granti*) is white, and it has exactly the same

FIGURE 4.1. Pale and dark morphs of the Parasitic Jaeger (*Stercorarius parasiticus*), also called Arctic Skua. (Art, Stephen Nash.).

DNA sequence as the white Red-footed Booby. This DNA sequence has undoubtedly been inherited by these two species from their common ancestor and underlies their white color. The white form of the Snow Goose has the same mutation at site 85 in this gene: it has a V instead of the blue form's M. The V mutation at site 85 in the *MC1R* gene, with the same effect on coloration, almost certainly happened independently of the boobies' mutation because these birds are very distantly related. As to why different color forms exist within populations of so many birds—that's in the next chapter.

Discrete polymorphisms are just the tip of the variability iceberg. Measure the bill, leg, metabolic rate, or just about anything else on a sample of birds, and you will find variation among individuals of the same species—just as in humans. This variation is more or less continuous: leg length doesn't come in separate small and large sizes, in either humans or herons. Both genetic differences among individuals and environmental effects that influenced their development can cause variation. In humans, height is inherited to some extent, but it can also be affected by nutrition during childhood. We can estimate the relative contribution of variable genes and variable environment on the principle that inherited genes cause offspring to resemble their parents, and relatives to resemble each other, in general. The greater the resemblance, the greater the effect of genetic inheritance relative to the effect of environment (with a caveat I'll raise in a moment). In fact, the proportion of variation that is caused by genes can be estimated by the correlation[6] between offspring and one or both parents. This proportion is called the *heritability* of the trait: a heritability of 0.9 means that 90% of the variation is caused by genetic inheritance and 10% by environmental effects. (I emphasize that heritability describes variation *among* individuals—0.9 heritability doesn't mean that an individual's feature is 90% due to genes and 10% due to environment.) All else being equal, the population average of a characteristic that affects fitness will change faster if its heritability is higher.

How can we find out if the variation in some characteristic is heritable? Research by Peter Boag, while studying with Peter Grant and Rosemary Grant, is a good example. Boag estimated the heritability of several characteristics in some of the Darwin's finches in the Galápagos Islands by measuring the traits of adults and their grown offspring.[7] He kept track of who was who by fitting each individual with a unique combination of colored leg bands (rings). In the Medium Ground Finch (*Geospiza fortis*), measurements of the wing, leg, and bill of the offspring and their parents were strongly correlated. The heritability of wing length was 0.91, and of the depth of the bill 0.90 (figure 4.2).

But was the offspring–parent resemblance necessarily due to genes? What if larger parents are better at getting food and raise larger, better fed offspring? What if some parents are larger because they grew up in places with better food and raised their offspring in those same places? Then an environmental influence on wing or bill development might be inherited as if by genes. (Such effects are familiar in humans: we tend to inherit our parents' language, religion, and socioeconomic milieu, with all of its many effects.) James Smith and André Dhondt answered this question for a population of Song Sparrows (*Melospiza melodia*).[8] They swapped twenty-four clutches of eggs and sixteen broods of newly hatched nestlings between nests and compared the features of the young birds, once grown, to those of both their true parents and their foster parents. The resemblance between offspring and true parents was nearly perfect (the heritability of beak depth was 0.98), whereas there was no

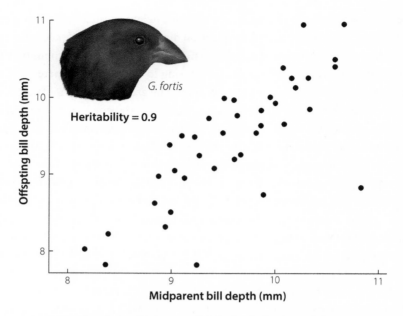

FIGURE 4.2. The relationship between bill depth of offspring and their parents in the Medium Ground Finch (*Geospiza fortis*). Each point shows the average bill depth in a family of offspring plotted against the average bill depth of their two parents. Similar studies were done in two years. The slope of the offspring versus parents' measurements corresponds to a heritability of 0.9, meaning that 90% of the variation is based on inherited differences and 10% on other factors, such as food or other environmental effects. (After Boag 1983; art, Luci Betti-Nash.)

correspondence at all between offspring and foster parents (the "pseudoheritability" was 0.01). It appears that most of the variation in quite a few physical traits of birds has a genetic basis.

The genetic component of continuous variation, whether it be in beak depth of a finch or human height, is usually caused by the presence in the population of two or more alleles (versions of a gene) at each of several or many genes, located at different places in the chromosomes. Each of these variable genes has an effect, often a very small effect, on the development of the characteristic. Think of each gene as having plus and minus variants (A^+, A^-) that add to or subtract a little from the feature. So my genotype might be $A^+A^-B^+B^-C^-C^-$, yours might be $A^+A^+B^+B^+C^+C^-$, and so your beak is bigger than mine. At least five different genes are known to affect beak depth in Zebra Finches (*Taeniopygia guttata*)[9] (but well over 200 genes affect variation in human height!).

So we know that there is variation and that it is strongly inherited. Is there any natural selection? Or is there evolutionary response to selection?

In chapter 1, I described how, in 1899, Hermon Bumpus published the first demonstration of natural selection on body size of House Sparrows that suffered mortality in a winter storm. A similar study was done by Charles and Mary Brown, who have studied colonies of Cliff Swallows (*Petrochelidon pyrrhonota*) in Nebraska for more than thirty years. In 1996, cold weather in late spring killed thousands of birds. Like Bumpus, the Browns measured many survivors and nonsurvivors and found that survivors were significantly larger, on average.[10] An interesting discovery was that the length of right and left wings was more symmetrical in survivors. The rare weather event, then, caused intense natural selection. It also caused an evolutionary change between the 1996 population and the offspring generation, measured one year later. Average body size was 1.7% greater in 1997 than in 1996 before the die-off. The change was much greater—huge, in fact—for wing asymmetry, which declined by 57%. These features are inherited, and the result of the mortality was natural selection that changed the population average. Of course, the change might not be permanent—under the more usual conditions, the larger, more symmetrical birds might not have an advantage—but the episode shows that considerable change can occur very rapidly.

This is the lesson, also, of the famous study of Darwin's finches by Peter Grant and Rosemary Grant, who are well known for their forty-year-long work on finch populations on one small island in the Galápagos archipelago.[11] They have documented selection and evolutionary change in several species of finches, especially in *Geospiza fortis*, the Medium Ground Finch. The most dramatic event was massive mortality in 1977, when a drought severely reduced the supply of the small seeds that were their major food. The survivors had larger than average bills that enabled them to feed on the large, hard seeds of a certain plant, *Tribulus* (figure 4.3). The average bill depth was about 8.8 millimeters before the drought but was 9.8 millimeters among the survivors (an 11% difference)—and this carried over into their offspring. But 1983 brought heavy rainfall that fostered the growth of vines that smothered *Tribulus* and led to an abundance of small-seeded plants. Again, there was massive mortality, but now small-billed birds, which feed more efficiently on small seeds, survived best. (The ability of finches with different bills to feed on different seeds was well documented by the Grants and others in earlier studies.) The result was a reversal of evolution, to a smaller average bill than the population had had in 1976.

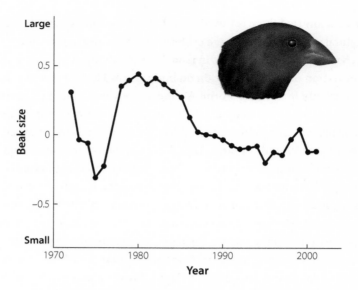

FIGURE 4.3. Changes in the average beak size in a population of the Medium Ground Finch (*Geospiza fortis*) over thirty years. The large decline, followed by increase in the 1970s and 1980s, shows how rapidly a genetically variable characteristic can change. "Size" is a composite of several measurements of the bill, with the changes shown relative to a baseline set at 0. (After Grant and Grant 2008; art, Luci Betti-Nash.)

A friend of mine who read about this work some years ago wasn't impressed. So the bill changed by about 1 millimeter, he said, and it didn't even stay changed—it reversed within a few years. What's the big deal?

Many great changes occur in small steps, both forward and backward; the stock market, for instance, is very much higher now than it was twenty years ago, but it didn't climb steadily: it zigged and zagged. Species such as certain clams and stickleback fish that have a detailed fossil record show similar, irregular changes that over time become considerable. And although a 1-millimeter change isn't much on a human scale, it means a lot to a small bird: it was an 11% change. A Marbled Godwit (*Limosa fedoa*), with a bill 120 millimeters long, is strikingly different from the much smaller Least Sandpiper, with an 18-millimeter bill—but at 10% per generation, this difference would take less than one hundred generations to evolve (plate 12). The Medium Ground Finch and the Cliff Swallow cases show that natural selection can be strong, that at least some characteristics have considerable heritable variation, and that a measureable evolutionary change can occur within one or a few generations. Can such slight changes be important?

Definitely. Biologists have described rapid adaptive evolution in a range of important characteristics. Here are three examples that I think are particularly interesting.

In 1939, a few House Finches (*Haemorhous mexicanus*)—a species that is naturally distributed throughout western North America—were released from a pet store in New York City. It took about fifteen years for the species to become noticeably abundant. (I remember, as a high school student, learning from older birders that the "Purple Finches" we were seeing were House Finches.) In 1994, birders first reported diseased House Finches, with swollen, infected eyes—a conjunctivitis caused by the bacterium *Mycoplasma gallosepticum* (MG). Within three years, the disease had spread widely and caused a 60% decline in the eastern population. But as the finches became scarcer, the prevalence of the disease declined, and by about 2007 it appeared that the population was becoming resistant. The finch population has recovered appreciably since then.

House Finches, like other birds and other vertebrates, have a dual immune system. The one we hear most about is the acquired immunity system (AIS), in which antibodies are produced that neutralize foreign proteins that the individual has already experienced in the past. (That is the basis of vaccination.) There is also an innate immunity system (IIS), involving cells and proteins that immediately respond to any proteins that aren't normally present on the surface of the organism's cells. This is the first line of defense, acting before the AIS can come into play. Many pathogenic microorganisms have evolved various ways of circumventing or neutralizing these defenses, and MG appears to suppress the AIS.

A team that included Scott Edwards, at Harvard University, and Geoffrey Hill, at Auburn University, showed that resistance in eastern House Finches has evolved by changes in the IIS.[12] In 2007, they experimentally infected birds from both the east (Alabama) and the west (Arizona), where there isn't any MG, and after fourteen days they assessed the level of activity of various genes involved in the two immune systems. They also assayed control birds that weren't infected. In infected Arizona birds, many of the genes—especially AIS genes—showed reduced activity compared with controls. The level of gene expression was similar to the level measured in eastern finches back in 2000, when they were dying in droves. The MG bacterium seems able to suppress those genes. But in the Alabama birds in 2007, none of the AIS immune genes were suppressed, and some genes in the IIS—the first line of defense—had higher activity in infected than in noninfected birds. The researchers concluded that the IIS had evolved in the eastern population and prevented the infection from suppressing the AIS, so both immune systems were active and together could vanquish the enemy. This is a case in which, very possibly, rapid evolution saved a population from extinction. Diseases such as malaria are thought to have caused the extinction of Hawaiian honeycreepers and other birds,[13] and of course the impact of infectious diseases has been terrifyingly evident in human pandemics such as the Black Death in the fourteenth century, the Spanish flu in the 1920s, and the 2020 coronavirus pandemic.[14]

Many bird species have changed their migratory behavior in recent times. The most extensively studied is the Blackcap (*Sylvia atricapilla*), a European warbler.[15] The number overwintering in the United Kingdom has increased dramatically in the last thirty years, owing mostly to birds that migrate northwestward from Germany and nearby areas in central Europe. This is a recent change in direction from the "standard" migration of German Blackcaps, toward the south and southwest.

Is this a case of evolutionary change? The answer depends on whether or not the difference in migration route is inherited. Peter Berthold compared the offspring of German birds that had overwintered in the United Kingdom versus southwestern Europe by means of a standard method for assessing birds' migratory orientation. Birds are placed in a funnel lined with typewriter correction paper[16] and are allowed to see the night sky, which they use to determine compass direction. They flutter up the slope of the funnel, leaving scratch marks on the paper that indicate their orientation. Sure enough, the offspring of birds that made the United Kingdom their winter home fluttered toward the west-northwest, while "standard" German birds oriented southwestward. Berthold then crossed UK-wintering adults with "standard" German birds and found that the orientation of their young was intermediate between their parents'—evidence that their orientation was genetically heritable. So there has been a partial genetic change—evolution—of the Blackcap's overwintering area and migration. Berthold suggested nine possible advantages of wintering in the United Kingdom, including less competition than in traditional Mediterranean quarters, earlier homeward migration in spring, milder winters, and the availability of bird feeders.

Thanks to citizen scientists, there is some evidence for the last two of these hypotheses.[17] The British Trust for Ornithology sponsors the Garden

BirdWatch program, in which more than 14,000 citizens record birds that visit their feeders. (A similar program in the United States, FeederWatch, is sponsored by the Cornell Laboratory of Ornithology.) Their data show that from 1999 to 2010, Blackcaps greatly increased their association with feeders and that they were particularly abundant in places with warmer winters.

Blackcaps also show less overall migratory activity[18] now than in years past. Each year from 1996 to 2002, Francisco Pulido and Peter Berthold captured and raised young birds and then reared their offspring, in turn. They exposed both the wild-caught parents and their offspring to the natural decline of day length in the autumn and measured their *Zugunruhe*[19] (migratory restlessness). The level of *Zugunruhe* declined by about 45% over the years, in each year's adults and in their offspring! A significant correlation between parents and their offspring indicated that about 43% of the variation was due to inheritance (heritability). For further evidence of heritability, Pulido and Berthold reared four generations of birds in the laboratory and imposed "artificial selection"[20] by rearing offspring only from birds that showed lower *Zugunruhe*. After four generations of such selection, the descendants of the original birds averaged 35% less migratory restlessness. Pulido and Berthold suggest that with the warming climate, the Blackcap might become a nonmigratory resident.

These and other evolutionary effects of climate change must be happening in many species. A twenty-four-year study of Blue Tits (*Cyanistes caeruleus*) in France showed that as spring temperatures have increased, birds that laid eggs earlier had more surviving offspring.[21] Will bird species evolve fast enough to survive the change in climate and other changes caused by human activity? I'll revisit this question in chapter 12.

Rapid evolutionary changes like these must be based on pre-existing genetic variation, not on new mutations, because new mutations happen so rarely. One of the cutest birds I have seen (scientifically speaking) is the Vinous-throated Parrotbill (*Sinosuthora webbiana*), which resembles a tit with a stubby bill that its name describes. The population in Taiwan, which ranges from sea level to over 2,000 meters, became isolated from the mainland population after the glaciers retreated and sea level rose.[22] Since then, lowland Taiwanese populations have experienced an increase of 9.6°C in mean annual temperature. In Taiwan, lowland populations now differ from highland populations in the DNA sequences of seventeen genes that affect oxygen use and regulation of body temperature and probably are adaptive in a hotter environment. Because the same DNA sequences are seen also in mainland samples, the advantageous mutations in these genes were apparently already present in the Taiwanese population when it started to adapt to higher temperatures.

I'm obviously a big fan of Darwinian evolution, and I wish everyone else were, too. Look at what it has wrought! But now I have to admit that some species can undergo important changes without any genetic evolution at all. Humans are the most conspicuous case: almost every aspect of our behavior, communication, and life differs radically from what it was just one hundred years ago. Some similar changes can be seen in birds, and they have two related causes: phenotypic plasticity and cultural inheritance.

In phenotypic plasticity, the same *genotype* (set of genes) can yield a different *phenotype* (the feature we see or measure) depending on the environment in which the individual develops.[23] The average waist-to-height ratio has increased in the United States in the last few generations because of processed food and soft drinks combined with little physical exercise. What you do with your muscles affects how big they are. A characteristic can be highly plastic in some species, and hardly at all in others: owls lay a variable number of eggs, depending on food supply, but hummingbirds almost always lay two eggs. So plasticity evolves: the degree of plasticity of a characteristic is determined to some degree by genes, and natural selection can favor genotypes that are more flexible in response to regularly recurring situations. But we are concerned here with the immediate effects of plasticity.

The Dark-eyed Junco (*Junco hyemalis*) typically nests at high latitudes or altitudes in North America, usually in coniferous forest. But in the 1980s, a population became established in the hot, dry environment of San Diego, on the coast of southern California. The San Diego population is sedentary rather than migratory, and it has a longer breeding season. A research team led by Ellen Ketterson, who had long studied the effects of hormones on this species' behavior and life history, and Trevor Price, an expert on bird speciation and diversity, compared the San Diego population and an ancestral-type montane population.[24] They found that compared with the "ancestor," the new population has less pronounced male plumage features (paler head, less white in the tail) and that males display less territorial aggression and less frequent extra-pair mating but greater care of offspring. These males have elevated testosterone levels for a longer period (matching the longer breeding season), but the level is lower than in the ancestral-type males. Birds with higher levels of testosterone have blacker heads, are more aggressive, and provide less care to their offspring. When recently fledged birds from both populations were kept for more than a year in an aviary, where they experienced the same environment, the birds did not differ in testosterone level.[25] The difference in tail color persisted, so it probably has a genetic basis, but the population's behavioral and life history adjustments to its new environment are an expression of phenotypic plasticity, mediated by flexible production of testosterone. This is a good example of how phenotypic plasticity, even without evolutionary change, can help a population survive changes in its environment.

Americans' change in diet is an example of *cultural evolution*,[26] which is based on imitation or learning from parents and often from other individuals. Oscine passerines learn many aspects of their song from parents or neighbors, and regional populations often diverge and form local song dialects. Some cultural "traditions" are certainly advantageous. In a famous example, Blue Tits and Great Tits in England learned to peck through the cap of milk bottles that were delivered to peoples' porches and eat the cream at the top, and the practice spread widely among tit populations.[27] Other innovations can spread just by cultural conformity, like fashion styles in society. Lucy Aplin and her colleagues caged some Great Tits with "puzzle boxes" that had two colored doors and trained some birds to obtain mealworms by sliding a blue door from left to right; they trained others to slide a red door from right to left.[28] Two blue-trained birds were released into one local population and two red-trained birds into a different population. Within a short time, about 75% of the birds in each population used the same solution as their educated "demonstrators." We do not yet know, overall, how important cultural evolution has been in adapting bird species to changes in their environment.

In cases like the House Finch and the Medium Ground Finch, researchers have studied evolutionary change in real time. But how can we judge if a feature is advantageous, and likely evolved by natural selection, if we have not seen the evolutionary change in progress? This can be very hard. In birds, as in other organisms, we see countless features whose adaptive advantage is a matter of debate or pure speculation, and we often don't know if these features have any advantage at all. I have no idea, and I can't find any published evidence (or even speculation!), on why the scales on the legs of most thrushes (Turdidae) and some other species are fused together, while they are separate in most other birds. In New Zealand I was delighted to see the Wrybill (*Anarhynchus frontalis*), a small plover that forages on shingle in river beds. Uniquely among birds, it has a strongly asymmetrical bill, bent sharply to the right (figure 4.4). Why on earth?

An observer of its feeding behavior wrote that the bill "was useful in capturing insects from the undersurface of stones where they would normally have

FIGURE 4.4. The Wrybill (*Anarhynchus frontalis*), showing its unique asymmetrical bill. (Art, Stephen Nash.)

been inaccessible to birds with shorter, straighter, or even upcurved bills." Wrybills also forage by plunging the beak into sand and sweeping sideways.[29] These behaviors suggest that the bill shape is an adaptation for feeding that evolved by natural selection in this special habitat. But they don't prove that the bill evolved by natural selection. Here is another possibility. We know that most genes affect more than one characteristic; so maybe the bent bill was an inescapable, even harmful, side effect of a gene that caused a very advantageous alteration of a different feature. Stuck with a weird bill, the species found a way of making the best of a bad situation and ended up with a good payoff. Now, I actually think this is really far-fetched, but it's not impossible. So how can we test for adaptive evolution due to natural selection?

Biologists have developed several approaches; I'll mention three. Probably the most reliable approach, when it is feasible, is to do experiments—alter some individuals or their environment and compare the effect with "control" individuals that were treated the same except for the alteration. Among the tanagers called flowerpiercers (*Diglossa*), some species feed on nectar by holding a tubular flower steady with the hooked tip of the upper bill and piercing a hole at the base of the flower with the chisel-like lower bill. Other species feed mostly on fruit and lack the hook. Jorge Schondube and Carlos Martínez del Rio tested the adaptive function of the hook by clipping the hook off the bill of a nectar-feeding species, so it resembled the bill of a fruit eater. (This did not injure the birds, which regrew their normal hooks as we regrow fingernails.) Compared with unmanipulated control birds, those with clipped bills improved their ability to eat fruit but were less able to pierce flowers and obtain nectar (figure 4.5). The hook is advantageous for nectar feeding but comes at the cost of efficiently processing fruit, the typical food of some other *Diglossa* species and of most other tanagers.[30]

A nonexperimental approach is taken in a whole field of biology called functional morphology, which uses principles of physics and engineering to demonstrate how a feature's form is suited to its function. For instance, the physics of aerodynamics is used to analyze how the shape and size of birds' wings determines their flying abilities.[31] The pressures that result from the flow of air over a wing create lift, which keeps a bird aloft, but the flow induces drag, especially along the rear edge and the tip of a wing. A greater wing span increases lift and is characteristic of many soaring species, such as vultures. Wings that are more pointed and slender induce less drag and are seen in fast-flying birds such as falcons and swifts. But that shape isn't as effective for maneuvering quickly, as in birds that must make rapid turns to fly through thick vegetation. These species, in many families, tend to have short, broad wings that can be flapped quickly and enable rapid acceleration. A similar approach, comparing species' features with what we think would be optimal design, is also used in studies of some behavioral traits. (Chapter 7 includes examples.)

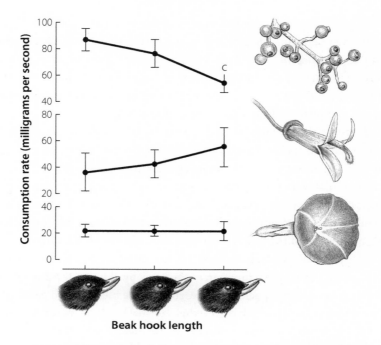

FIGURE 4.5. This graph shows the average rate of consumption of small fruits (top) and nectar from two kinds of flowers by Cinnamon-bellied Flowerpiercers (*Diglossa baritula*) with a normal hook (silhouette at bottom right) or with a partly or fully shortened hook (left). Birds with shortened hooks fed less effectively on *Ipomoea* flowers but more efficiently on fruits. (After Schondube and Martínez del Rio 2003; art, Stephen Nash.)

A third way that evolutionary biologists test ideas about whether and how a feature is adaptive is to examine patterns of convergent evolution, comparing different species to see if a similar feature has repeatedly evolved and is correlated with the way it is used or with the species' habitat or way of life. Bills have evolved to be finely pointed in many groups that glean insects, deep and more or less conical in groups that eat seeds, and similarly hooked in hawks, falcons, owls, and shrikes.

This approach to studying adaptation is called the comparative method. If we see that the same feature has evolved again and again in a similar context in different bird lineages, the correlation provides evidence that the feature is advantageous in that repeated context. A correlation requires multiple instances, and they have to be independent of each other to count as independent

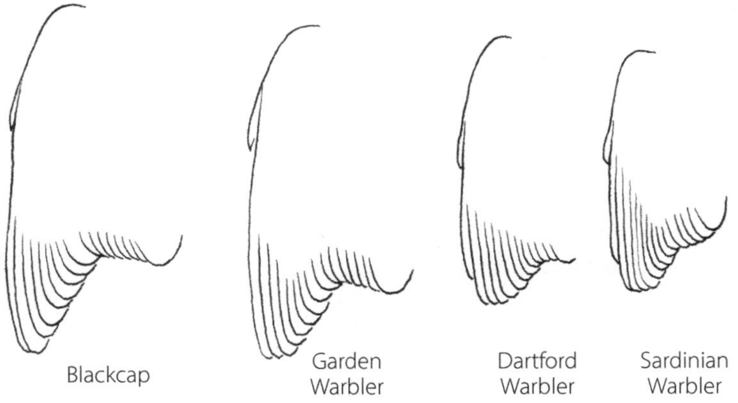

Blackcap   Garden   Dartford   Sardinian
            Warbler   Warbler   Warbler

FIGURE 4.6. The wings of four species of *Sylvia* warblers. From left to right, the outer primaries are longer in the two highly migratory species, the Blackcap (*S. atricapilla*) and Garden Warbler (*S. borin*), than in the more sedentary Dartford Warbler (*S. undata*) and Sardinian Warbler (*S. melanocephala*). (After Kipp 1942; Rensch 1947; art, Stephen Nash.)

evidence. This is where phylogeny comes in, by showing that a feature has evolved independently, convergently, in multiple lineages. This was clear to one of the giants of twentieth-century evolutionary biology, Bernhard Rensch, whose *Evolution above the Species Level* was published in German in 1947 and in English in 1959.[32] To illustrate that pointed wings are advantageous for birds that take long flights, he described the wings of four species of *Sylvia* warblers: two nonmigratory species with blunt wings and two migratory species (Blackcap and Garden Warbler [*Sylvia borin*]) with long, pointed wings (figure 4.6). (This feature, referred to as "primary extension," is used by birders today to help distinguish some similar species, such as *Acrocephalus* warblers in Europe and *Empidonax* flycatchers in North America.) Suppose the Blackcap and Garden Warblers were closest relatives and had inherited the pointed wing from their common ancestor. Then they would provide only one case of a pointed wing evolving in migratory species, not two: we would have less evidence than we thought. For this reason, Rensch stressed that the same difference in wing shape has evolved repeatedly in distantly related groups of birds: "the same holds good when we compare [migratory and nonmigratory] species of wheatears, redstarts, orioles, and the like." Rensch's conclusion has been confirmed by many recent phylogenetic studies; for example, wing shape is correlated with migration distance in different lineages of shorebirds.[33]

There is no doubt that many—maybe most—of birds' characteristics affect survival or reproduction and have evolved by natural selection. But can we say all? No. Several evolutionary factors besides natural selection might explain some characteristics. And even if natural selection is involved, figuring out just what causes the selection and how it determines trait evolution can be very challenging. Some of the best examples of cases solved and unsolved—and some fascinating stories—have come from biologists' efforts to explain the variation within species. That is the topic of the following chapter.

5

The Ruff and the Cuckoo

VARIATION WITHIN SPECIES

Evolutionary biologists are in love with variation. We are obsessed with it, dream about it, talk endlessly about it to anyone who will listen. Why? First, because variation among species—the immense diversity of species and their characteristics—is exactly what we want to explain, exactly what motivates us. Second, because variation within species is the raw material, the *sine qua non*, of evolution. Almost everything we know about how evolution happens is based on studies of variation within and among species. Biologists who asked why there are different color forms of Ruffs (*Calidris pugnax*) and White-throated Sparrows have clarified or illustrated fundamental ideas about evolutionary processes—and have shown just how fascinating these species are.

When we see variation among individuals of a species, we can first ask if it is caused by differences in their genes, in the environment they have experienced, or both. Eastern populations of the Northern Flicker (*Colaptes auratus*) have yellow flight feathers (the "yellow-shafted" subspecies), while those in western North America have red ("red-shafted"). These forms interbreed in central North America, where studies have shown that the color difference is inherited: it has a genetic basis. In eastern North America, far from the hybrid zone, occasional "yellow-shafted" flickers are seen that have some red or orange flight feathers—but no other features of the western race. Their color is based on the carotenoid pigment rhodoxanthin, which they have acquired by eating the fruits of Asian species of honeysuckles that are planted as ornamental shrubs.[1] The same explanation accounts for red (rather than the normal orange) color of some Baltimore Orioles and for the red (rather than the normal yellow) tail tips of occasional Cedar Waxwings (*Bombycilla cedrorum*). The normal yellow and red colors found in most birds are carotenoid pigments that are obtained in their diet (so variation may be environmental rather than genetic). But genetically encoded proteins determine where these pigments are deposited in growing feathers, and in some cases these proteins modify the carotenoid molecules. (That's why the difference between yellow-shafted and red-shafted populations of the flicker is genetic.) So variation in color can have both an inherited component and an environmental component that arises from the experiences of individuals during their development.

It is the genetic variation that is the basis of evolutionary change and where we confront some intriguing questions. Consider *polymorphic* species, such as the pale and dark morphs, or "phases," of the Parasitic Jaeger (figure 4.1) or the white and "blue" morphs of the Snow Goose. If the morphs occupied different geographic regions, we might reasonably expect that they are in different environments that favor different colors for some reason. But if the morphs coexist, we can ask, "Why hasn't one completely replaced the other?" Surely one of the morphs must have a higher rate of survival or reproduction. And even if the morphs have exactly equal fitness, we would expect one to ultimately replace the other by the process of genetic drift. Something must be maintaining the variation. This chapter describes some causes of variation within bird populations, drawing on studies of several species that each tells an interesting story.

Genetic variation exists, first of all, because genes mutate. But this doesn't explain the coexistence of two or more genetic morphs because any specific gene mutation, such as one for darker coloration, arises very rarely. By comparing DNA sequences of Collared Flycatcher (*Ficedula albicollis*) parents and offspring,[2] researchers found small numbers of new mutations in the offspring and calculated that the rate at which an average nucleotide base pair mutates is 4.6 mutations per 100,000,000 sperm or eggs, or 4.6×10^{-9}. This is the rate at which a single new copy of a particular mutation (a specific allele of the gene) would be added to the population per generation. (It is roughly the same in mice but higher in humans.) If a population has, on average, N breeding individuals every generation, most genes in the genome are represented by $2N$ copies (since each individual receives one gene copy from each parent). So if a specific new mutation starts out with one copy, its frequency (its proportion of all gene copies) is $1/(2N)$: say 1 out of 2 million if the population size is 1 million individuals. Every time this mutation occurs again, the frequency of that allele is increased, but this increase is *very* slight if the mutation arises only a few times per 10^9 eggs or sperm. So repeated mutation, by itself, is unlikely to cause a population to have two or more common morphs.

How, then, can the frequency of the mutant allele increase substantially within a population?

There are three ways: gene flow, genetic drift, and natural selection. An example of the gene flow hypothesis is the Snow Goose in eastern Canada, where almost the entire population was once composed of the white morph. In recent decades, the proportion of "blue" geese increased by incursion from the west, where the "blue" form is more common.[3]

Gene flow is unlikely to explain cases, such as the Parasitic Jaeger, in which birds are polymorphic throughout the species' range. The obvious solution, we might say, is that both alleles or colors improve some aspect of fitness. Perhaps there are situations in which each is advantageous, so that neither one can completely take over?

That is a hypothesis that most biologists entertain, and may try to test, when they see a polymorphism. But we will have to rule out one more evolutionary process: random genetic drift (see figure 1.3). This is purely random fluctuation in the frequencies of two (or more) alleles, from generation to generation. If a population has, say, 10% dark and 90% pale individuals, you are likely to see 9% and 91%, or 12% and 88%, or some other slight deviation from 10:90, in the next generation because some potential parents died, or had slightly fewer or more offspring, purely by chance, not because of their genotype. Some were luckier than others, who were hit by cars or failed to find mates or had fewer offspring for other reasons that had nothing to do with their coloration or their genes. The mathematics of probability tells us that if these random deviations, sometime up and sometimes down, continue through the generations, one allele will eventually zigzag to fixation (100% of the gene copies) or will be lost (0%). A very rare allele (such as one that recently arose by mutation) has a small probability that it will eventually increase and be fixed at frequency 1.0 (100% of gene copies). Before fixation or loss occurs, both alleles have intermediate frequencies—which might possibly account for the polymorphism we see. But eventually one or the other allele will be fixed, so a long-lasting polymorphism must be caused by some kind of stabilizing process.

Genetic drift is most important if a mutation is *neutral*, only slightly affecting the organism's fitness or not at all. These include some mutations that are synonymous: they don't alter the amino acid in the encoded protein and so they have little or no effect on the organism. These mutations can float up to high frequency in a population purely by genetic drift—and every population is full of them. In just about every species, the genome is full of single-nucleotide polymorphisms (or SNPs, pronounced "snips"), most of which are synonymous: some gene copies have different base pairs at the site but no difference in the encoded protein.

These SNPs are immensely useful in genetic studies. They are the chief basis for identifying people by their DNA and for determining who was or wasn't the father in paternity lawsuits (or which bird was some nestlings' father—it's not always the female's "social mate"). They can be used to estimate how large a population has been in the relatively recent past: because mutations are either lost or fixed faster in small than large populations, a large population will have more SNP variation. (Among many examples, the Zebra Finch colonized the Lesser Sunda Islands from Australia about a million years ago and has only about 20% as much DNA sequence variation in the islands as the much larger Australian population.[4]) Biologists use SNPs as "genetic markers," or landmarks, when they try to find out where a gene is located in the genome. They might look for genes that affect bill size by sequencing the genome of many individuals of a species. If they find that alternative base pairs of a certain SNP, known to be located in a certain place in the genome, are consistently correlated with bill size, they suspect that a gene is nearby that affects bill development.[5]

Might visible bird polymorphisms be like neutral SNPs and result from genetic drift? Possibly, and this would be the "fallback" hypothesis that a biologist would accept if no evidence can be found for natural selection or any other hypothesis. We do have evidence that genetic drift can affect small populations. One example is egg hatching rates of birds in New Zealand.[6] Habitat destruction and persecution have reduced the population size of many native species to much lower levels in some species than others. Also, many species of nonnative birds have been introduced into New Zealand, some of which started from smaller numbers than others. Among twenty-two native species, the fraction of eggs that fail to hatch is much higher in species that have gone through a period of lower population size. And among fifteen introduced species, the smaller the number of birds that were introduced (mostly from Europe), the lower the hatching rate. Why? At a great many genes in the genome, mutations occur that are harmful. Many of these persist within populations because although natural selection reduces their frequency, they continually recur. The incidence of leucism (whitish coloration), for instance, is about 1 in 20,000 Gentoo Penguins (*Pygoscelis papua*) and 1 in 114,000 Adelie Penguins (*Pygoscelis adeliae*).[7] These mutations may increase by genetic drift if the population is small. In small populations, many individuals are closely related, and so they will tend to carry the same genes, some of which are likely to be harmful mutations. If two carriers mate, their offspring are likely to be have two copies of the same harmful gene and be severely affected. They suffer *inbreeding depression*. This is the cause of low hatching rates in some New Zealand birds. In the highly endangered Kakapo, a giant, flightless, nocturnal parrot (total population 148 in 2018; plate 5), more than 60% of eggs fail to hatch.[8] Even in the Collared Flycatcher in Europe, which has a large population, there are so many rare, harmful mutations that among offspring from closely related parents, the number that survive to breed is reduced by 94%.[9]

Bearing in mind the possibility of genetic drift, might some kind of natural selection explain polymorphic variation? Yes, but I first have to mention an indirect, rather than direct, effect of natural selection. A single chromosome has thousands of genes along its length. So if an advantageous mutation (say, 3*) happens in gene number 3 on a single chromosome in a single individual, and that mutation increases the number of descendants so that the frequency of the mutation increases in the population, it will drag along nearby (*linked*) genes on that chromosome, with whatever rare mutations those genes may include. So the many descendants of the bird with the 3* mutation may also carry mutations in genes 9, 55, and so on, that themselves may be neutral or at least not too harmful. This is called *genetic hitchhiking*, and it could give the appearance that an allele is advantageous even though it has become common just by hitchhiking with an advantageous mutation in a different gene. Later in this chapter, I will describe several species, such as the Ruff (figure 5.1), in which genes that underlie color variation are inherited together with many other genes so that a change in the frequency of one gene, owing to natural selection, would cause others to change in concert.

Back to polymorphism: what explains the cases of two or more common genetic types within populations? Researchers have found several ways in which natural selection can keep a population polymorphic.

One possibility is that each genotype has superior fitness in one or another aspect of a variable environment: perhaps each is better adapted to a different habitat or type of food. The best example I know of is the Black-bellied Seedcracker that I described at the start of chapter 1, in which small-billed and large-billed individuals feed most efficiently on different seeds and have better survival than birds with intermediate bills.

Might the two favored morphs eventually become different species? Possibly. An intriguing case is the Red Crossbill (*Loxia curvirostra*) complex in North America, which includes several "call types" that differ in bill structure

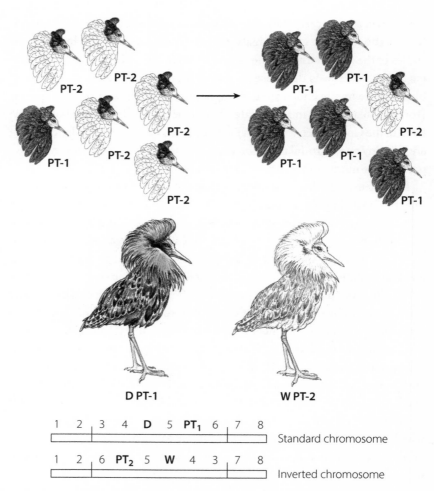

FIGURE 5.1. Genetic hitchhiking, using the Ruff (*Calidris pugnax*) as a hypothetical example. Below, dark and white male Ruffs, with associated diagrams of their chromosome genotypes. Dark males have two copies of the "standard" chromosome. Among its genes, numbered 1 through 8, is a gene marked D that affects coloration of the male's ornamental ruff and another gene, PT, that encodes palmitoyl transferase, a steroid hormone receptor protein. Dark males have the D allele of the color gene and the allele PT_1 of the receptor gene. White-ruffed males have one copy of the standard chromosome and one copy of a chromosome with the allele W of the color gene and a different allele, PT_2, that encodes an altered receptor protein. This chromosome has an inverted segment: the order of genes from 3 to 6 is reversed, including both the color and the receptor genes. PT_1 is completely associated with the D color allele, and PT_2 with the W color allele, because the inversion prevents interchange between the genes. In the diagram above, we suppose dark (D) males (composing 17% of males in the population, at left) have a higher reproductive rate than white (W) males and increase in frequency (to 83%, at right). That would cause an increase in the frequency of the PT_1 allele, which "hitch-hikes" with D. (Art, Stephen Nash.)

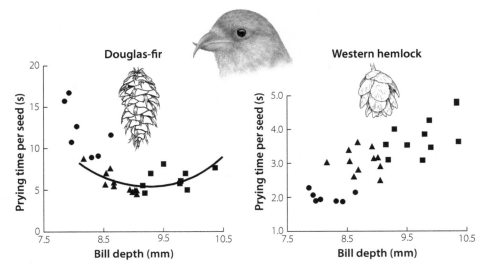

FIGURE 5.2. Three "call types" of Red Crossbill (*Loxia curvirostra*), shown as circle, triangles, and squares, differ in bill size (depth) and in the efficiency with which they extract seeds from different cones. (Right) The small cones of western hemlock are most efficiently used by call type 3; each circle is data from one individual bird. (Left) Call type 3 feeds very inefficiently on the larger cones of Douglas-fir, which call types 4 and 2 handle better. (After Benkman 1993; art, Luci Betti-Nash.)

and more efficiently extract seeds from the cones of the different conifers on which they specialize (figure 5.2). Birds with like bill morphology tend to mate together, using differences in their flight calls to recognize suitable mates. One such population has been recognized as a distinct species.[10] (I will revisit this theme in chapter 10, on speciation.)

It might sound like a stretch, but the two sexes may be thought of as different "environments" in which different alleles of a gene might increase fitness. Two alleles may persist if each allele enhances fitness of one sex but diminishes fitness of the other sex.[11] A likely example is wing length in the Great Reed Warbler (*Acrocephalus arundinaceus*).[12] Wing length is correlated between related males and females (e.g., siblings), evidence that some of the same genes affect wing length in both sexes. In a twenty-year study of color-banded birds in a Swedish population, investigators estimated the lifetime reproductive success (fitness) of every breeding individual. They found that fitness is higher for longer-winged males but for shorter-winged females. So alleles for both longer and shorter wings would persist because each kind of allele is propagated by one or the other sex. Longer wings are known to be advantageous for migrant species (chapter 4). Longer-winged male Great Reed Warblers arrive earlier in the spring, often have several mates, and have high reproductive success. It's possible that short wings are advantageous to females in maneuvering within the dense stands of reeds in which they nest. Whatever the reason, "sexually antagonistic selection" maintains variation.

Most of the known polymorphisms in birds involve conspicuous differences in coloration that are familiar to many birders. It isn't easy to find out just what keeps this variation stable because the genetic basis of coloration is usually complex, and many of the genes involved affect not only color but other features as well. For example, the *MC1R* (*melanocortin-1-receptor*) gene encodes a protein that binds hormones (melanocortins) that stimulate production of the dark pigment melanin, but the gene can affect immune responses and other physiological functions as well. (Pleiotropy is the technical term for a gene's affecting more than one characteristic.) And coloration can play many roles and have diverse costs and benefits. The melanin pigments that make for dark coloration can affect temperature balance (dark absorbs more sunlight), feather wear (melanin reduces degradation of feathers by bacteria), cryptic match to a bird's background, and social interactions, including with mates.[13]

My first example is a gorgeous bird found in northern Australia, the Gouldian Finch (*Erythrura gouldiae*)—so gorgeous that it suffered greatly from trapping for the cage bird trade until at least 1981, when trapping became illegal. You can buy them, though, because they are bred in captivity. Most birds, of both sexes, have either a red or black face, a difference caused by a single gene; black birds make up about 70% of wild populations (plate 13). In a comprehensive series of studies, Sarah Pryke solved the problem of polymorphism in this wonderful bird.[14] Reds tend to mate with other reds, and blacks with other blacks. This is advantageous because the offspring of mixed pairs have very high mortality. Reds are much more aggressive and compete more successfully for high-quality tree holes in which they nest (and for nest boxes in the experimental aviaries that Pryke used). As long as reds aren't too common, they produce more offspring than blacks, on average. But their aggressiveness comes at a cost that the more placid blacks largely avoid. When they spend a lot of time being aggressive, they suffer stress, have lowered immune function, spend less time caring for the offspring, and have fewer surviving offspring—which, moreover, are smaller and less able to mount strong immune responses. There is a trade-off between the costs and benefits of competitive aggression versus parenting. As the proportion of red morphs in the population increases from generation to generation, they have more and more aggressive encounters and their fitness becomes lower than that of the black-faced birds, which rear more and healthier offspring. But as black-faced birds become more prevalent, reds spend less time fighting each other, so they are less stressed and rear more offspring, which successfully compete against blacks. The Gouldian Finch is a fine example of "negative frequency-dependent selection," meaning that the higher the frequency (percentage) of a genotype in a population becomes, the more its fitness drops, to the point that the average fitness of individuals with the majority genotype falls below that of the minority genotype. The minority genotype then increases, but as it becomes more common in turn, it loses its advantage. So no one genotype can take over the population.

A curious aspect of this story is that reproductive success isn't determined directly by the color difference: it is caused by behaviors that are *correlated* with color and that affect hormonal and immunological status and success in competition. The color seems to serve as an indicator of how the individual behaves, and indeed, Pryke showed that a bird will react to a black-faced bird fitted with a red mask as if it expects that bird to be aggressive. But surely we would expect facial pigmentation and the brain circuits that produce behavior to be governed by different genes. How are the presumed genes for behavior and coloration held together, since we would expect them to be separated by recombination? (Remember, chromosomes with different alleles of various genes swap parts.) The answer, for the Gouldian Finch, is somewhat more complicated than we need to go into. It's simpler in another fascinating case, the White-throated Sparrow.

The White-throated Sparrow is familiar to most birders in eastern North America. About half of the population, of both sexes, have sharply contrasting white and black head stripes and half have tan and duller black stripes (plate 14). The remarkable feature, first reported in 1961 by James Lowther, is that almost all mating pairs are a combination of a white-striped (WS) bird, of either sex, and a tan-striped (TS) mate. This probably results from the strong behavioral differences between the morphs, paralleling the behavior of the Gouldian Finch morphs: WS birds, especially males, are much more aggressive than TS birds: they defend territories effectively and seek adulterous "extra-pair copulations" more actively than TS males. Female WSs also contribute to territory defense. But TS males are better parents and spend more time guarding their mates against the adulterous WS males.[15]

The differences in color and behavior are traceable to genes that are inherited together as a unit because they lie within a polymorphic structural variation of a chromosome, called a chromosome inversion. Very occasionally, a chromosome with its many genes will loop back on itself, break in the two sites of contact, and reattach the broken ends, so that the genes within the loop now are in reverse order. So if we have genes labeled 1 through 16 along the chromosome, and the loop makes contact near genes 5 and 12, the newly repaired inverted chromosome has the sequence 1-2-3-4-5-**12-11-10-9-8-7-6**-13-14-15-16. If an individual is heterozygous for the inversion—if it inherited an ancestral-type "standard" chromosome from one parent and an inverted chromosome from the other parent—the genes within the inverted region can't be swapped between the two chromosomes. So if several genes within this region have different alleles in standard and inverted chromosomes, those several genetic differences are inherited as a unit. That is exactly the case with one of the sparrow's chromosomes, which has an inversion that contains about 1,000 genes. One of these genes controls stripe color, with alleles that we can call w and t. Tan-striped birds are homozygous for the standard chromosome, which has the t allele, so they are genotype tt. White-striped birds are heterozygous, with one standard and one inverted chromosome, which has the w allele, so their genotype is wt. The wt (white-striped) birds produce equal numbers of w-carrying and t-carrying sperm or eggs, which unite with the t-carrying eggs or sperm produced by their TS mates. The result is equal numbers of wt and tt offspring. Elementary genetics explains why there are equal numbers of the two morphs in White-throated Sparrow populations.

But equal numbers occur only because the behavioral differences in aggression and parental care lead to "opposite mating" between the two morphs. This is certainly advantageous—it combines good territorial defense by one individual and good parenting by the other. What causes those behavioral differences? One of the crucial differences is in the level of a protein, ERα, in certain parts of the brain, due to a difference in a part of the *ER*α gene that regulates where and how actively it is transcribed into RNA, which in turn is translated into protein.[16] This protein, in the membrane of brain cells, binds certain steroid hormones, which thereby can affect behavior. White-striped birds have much higher *ER*α activity in the medial amygdala, a part of the brain that processes sensory information related to social behavior, including aggression. The higher parental care by tan-striped males is correlated with higher *ER*α activity in another brain region that is known to affect parental behavior in birds. I think it is so cool that a familiar bird can be the nexus of evolutionary biology, behavioral biology, brain science, and genomics, working together to explain observations by curious naturalists.

Most shorebirds have fascinating behavior, none more so than the Ruff, named until recently *Philomachus pugnax*: the pugnacious lover of battle. (Alas, phylogenetic study has now moved the species into the more prosaically named genus *Calidris*, along with stints, knots, and many other sandpipers.) Ruffs mate in leks: aggregations of males, each of which aggressively defends a small mating territory within a larger arena where females come to mate. The breeding male Ruff is spectacular, with head plumes and a collar of long plumes over the breast. These ornaments are very variable: they may be uniformly black, rufous, white, or a mixture (figure 5.1). The most interesting variation among Ruffs is that while dark-ornamented "independent" males defend their territories, white-ornamented males, referred to as "satellites," don't defend space. They enter independent males' territories and exhibit submissive behavior, to which the territory owner responds by squatting in a posture used when courting females.[17] (There is also a third, much rarer male morph, called "faeder," with female plumage and similarly submissive behavior.)

The most remarkable part of this story is that independent males are more successful in mating if they have satellite companions, which apparently are especially attractive to females; in fact, the two males display jointly. The satellites also get to reproduce (without paying the cost of constantly defending a territory against pugnacious males, which physically attack each other). Female Ruffs (called reeves), more often than not, mate with and have offspring by two or more males and mate with both independent and satellite males more often than by chance. Males, then, have two different reproductive "strategies," territorial and "sneaker," which are known also in diverse other animals. According to a mathematical model, the polymorphism in this species may persist indefinitely because independent and satellite males mutually benefit each other.

And to add a genetic note, the genes that affect the variable coloration and behavior are locked together in a chromosome inversion, much like the situation in the White-throated Sparrow. Independent males have ancestral, noninverted chromosomes, while satellites have one copy of the inverted version and one noninverted chromosome.[18] The chromosomes differ in a number of genes that affect secretion and binding of steroid hormones. No Ruffs are homozygous, with two inverted chromosomes, because one of the break points where the inversion was formed broke an important gene. Any Ruff embryos that inherit two inverted chromosomes have two damaged, nonfunctional copies of that gene, so they die before hatching. This situation is reminiscent of human sickle-cell disease. In West African populations, people with a single copy of a mutated hemoglobin gene (the sickle-cell allele) are more resistant to malaria than people with only "normal" hemoglobin. But the mutation reduces hemoglobin function, and homozygotes, with two copies of the mutation, suffer sickle-cell disease and typically die at a relatively early age. Adaptation to malaria comes with a dreadful cost. Natural selection isn't a caring, provident Mother Nature; it is only statistically consistent differences in survival and reproduction, no more.

The polymorphism that may be most familiar to birders is the contrast between darker and paler or reddish versus grayish, as seen in Rough-legged, Short-tailed (*Buteo brachyurus*), and Red-tailed Hawks (*B. jamaicensis*), Common Buzzards (*B. buteo*), jaegers, and many species of owls. In some cases, the color forms definitely differ in fitness, but it has been hard to pin down exactly why the polymorphism persists. I mentioned earlier that melanin pigments can have diverse roles, affecting body temperature, feather wear, social interactions, and cryptic patterns, and that a melanin-producing gene also affects immune responses and other physiological functions.[19]

A survey of the world's owls classified 69 of 206 species as polymorphic for color, usually with rufous and gray forms.[20] The authors reported that polymorphic species are strongly associated with habitats that have "intermediate" vegetation cover, neither fully open nor densely forested. They suggest that the color forms may each be advantageous in different (light or dark) habitats and may be maintained by variable natural selection. But the many studies of Tawny Owls paint what appears to be a large, more complicated picture.[21] This species has dark and light rufous morphs that differ in their feathers' melanin content, as well as a gray morph that lacks melanin (plate 15). One study, in Italy, found that rufous birds have more parasites than gray birds during the breeding season and reproduced less successfully. A study in Finland found that the offspring of gray birds had higher survival, possibly due to better parental care. And the environment may play a role: the darker brown form has increased in Finland over a forty-eight-year period, correlated with warmer winters. This morph suffers high mortality during winters with heavy snow, but such winters are becoming rarer. Somehow, the advantages stemming from these various differences probably maintain the genetic differences that affect coloration, but the picture isn't yet complete.

The picture is also incomplete, but tantalizing, for some hawks. In South Africa, the Black Sparrowhawk (*Accipiter melanoleucus*) has white- and black-bellied forms that Petra Sumasgutner and colleagues studied for eight years.[22] They found that the offspring of mixed pairs survived better than offspring from like pairs. The authors speculate that light and dark parents might forage in different (lighter or darker) habitats and bring more food to the nest than like pairs, but so far evidence is lacking. Common Buzzards range from quite dark to quite light, and two research teams, in Germany and the Netherlands, reported that intermediate birds produced more surviving offspring.[23] The reasons are not known, but one factor that might contribute is that compared with intermediate birds, the light and dark morphs each carry higher loads of a different parasite.[24]

My final example of genetic polymorphism is part of a larger, fascinating topic: brood parasitism.[25] This is familiar to North American birders in the Brown-headed Cowbird (*Molothrus ater*), to Europeans in the Common Cuckoo (*Cuculus canorus*), to Australians in the bronze cuckoos and the huge Channel-billed Cuckoo (*Scythrops novaehollandiae*), which parasitizes crows.[26] The habit of laying eggs exclusively in the nests of other species (hosts) has also evolved three other times: in the honeyguides, one duck, and the Viduidae—African finches that include the widowbirds, whydahs, and the anomalous Cuckoo-finch (*Anomalospiza imberbis*). Parasite nestlings compete with the host's nestlings for food and, in some cases, push the host's eggs or young out of the nest or kill them outright. Not surprisingly, many hosts have evolved various defenses, such as mobbing or attacking adult parasites and removing parasite eggs from their nest. Oddly, most host species do not reject or recognize parasite nestlings, even when they are grossly different from the host's own young.[27]

The Common Cuckoo has been studied more than any other species, especially by Nick Davies and his collaborators at Cambridge University. The most fascinating feature of the Common Cuckoo is egg polymorphism. A population of cuckoos includes several different genetic types of females (called gentes, Latin for "tribes"; singular, gens). Each gens parasitizes a different host species and lays eggs that closely resemble the host's. The gentes differ in genes that affect egg color and in maternally inherited mitochondrial DNA but are all the same species. (Claire Spottiswoode, another leading researcher on brood parasites, found a similar pattern in the Greater Honeyguide [*Indicator indicator*]. The mitochondrial DNA distinguishes two gentes that parasitize species that nest in tree holes versus burrows in the ground. The sequence difference in the mitochondrial DNA is so great that these lineages may have persisted for 3 million years.)[28]

For at least five decades, biologists have used artificial eggs to study how host species of birds react to brood parasites' eggs.[29] Michael Brooke and Nick Davies added to the nests of host species model eggs that in size, shape, weight, and especially coloration closely resembled the mimetic eggs that each Common Cuckoo gens typically lays (plate 16). By ejecting the egg or deserting the nest, most host species rejected unlike models more often than like models, providing clear evidence that a cuckoo genotype's fitness depends on how closely its eggs resemble those of the host.[30] Species such as the Reed Warbler (*Acrocephalus scirpaceus*), which have a long history of parasitism by cuckoos, tend to reject eggs that differ from their own in color or the pattern of markings, whereas species north of the cuckoo's distribution do not. This difference indicates that the ability of most host species to recognize and reject cuckoo eggs is a result of natural selection by cuckoo parasitism.

Another way hosts can foil egg-mimicking parasites is to evolve greater variation among eggs,[31] so that the parasite's eggs are less likely to match and can be rejected. In two families of African birds, those species that are parasitized have more variable combinations of egg color and markings than non-parasitized species. The eggs of the Tawny-flanked Prinia (*Prinia subflava*), a warbler-like host of the Cuckoo-finch, vary in ground color (white, blue, brick red, olive green) and in the intensity, size, and spatial pattern of darker markings. The eggs of the Cuckoo-finch vary in the same ways. Claire Spottiswoode and Martin Stevens moved prinia eggs between nests in order to determine which differences between the owner's eggs and the foreign egg would result in egg rejection. They concluded that the birds integrate all these features, especially color, as cues to recognize foreign eggs. These authors then compared eggs from their field study, in 2007–2009, with a collection of eggs made mostly in the 1970s and 1980s. They found that in both species, variation in egg features increased over time (so both species were evolving) and that parasite eggs matched host eggs better in the same time period than if compared between the different times. That is, the changes in the parasite paralleled changes in the host. These results are just what we would predict if there is an ongoing evolutionary "race."

Does either species ever "win"? Manuel Soler, who has studied a large cuckoo that parasitizes magpies, notes that many species that look like promising hosts are not parasitized but nevertheless reject foreign eggs.[32] He suggests that they may once have been parasitized but became so proficient at recognizing parasite eggs that the parasite abandoned them and shifted to more susceptible host species. This may have happened in Hungary, where, since the

1960s, Common Cuckoos seem no longer to parasitize Red-backed Shrikes (*Lanius collurio*).[33] Whatever the outcome of the coevolution between parasites and hosts may be, they dramatically show how inherited variation is at the heart of evolution.

———

The models of how variation might persist were developed by evolutionary biologists who might not have known a hawk from a handsaw but were adept in mathematics. These models, which I have explained verbally, show that the variation we see has a limited set of possible explanations. There might be mixture (gene flow) of two spatially distinct populations that are each adapted to a different environment (the Snow Goose is a likely example); adaptation of different genotypes within a single population to different parts of the species' ecological niche (as in the Black-bellied Seedcracker and the Common Cuckoo's host-adapted gentes); higher fitness of rare genotypes than common genotypes (as in the Gouldian Finch and perhaps the Ruff); selection for different alleles in females versus males (probably the Great Reed Warbler); or higher fitness of heterozygotes (with two different alleles) than homozygotes (White-throated Sparrow).

All of the examples I've described seem to fit one or more of the theoretical models. But in almost all cases, close study of a bird species has revealed something special and unexpected, some delightful quirk of behavior or genetics that excites our sense of wonder, enlarges biology beyond the abstract realm of theoretical expectations, enriches our appreciation of the individuality of each species and of the extraordinary diversity of birds and of living beings as a whole. Who would have expected a sparrow to mate only with its alternative counterpart? Isn't it remarkable that a sandpiper cheats in the mating game by carrying a white flag, inducing competing males to tolerate him, even to treat him with the deference due a female? Isn't it interesting that Common Cuckoos differ in the color of their eggs and know which host species to parasitize? Among more than 10,000 species of birds, how many other surprises await the curious naturalist who asks why?

6

Hoatzin and Hummingbirds

HOW ADAPTATIONS EVOLVE

In 1990, I had the privilege of joining an expedition up the Amazon River in eastern Peru that was led by Terry Erwin, a curator of insects at the Smithsonian Institution's National Museum of Natural History. The main aim of the trip was to collect canopy-dwelling insects by shooting a fog of insecticide into a tree's canopy. Within minutes, hundreds of insects would drop onto sheets placed below the tree, including far more species of beetles (Terry's specialty) than anyone had expected before. Of course, we saw plenty of birds, most impressively when the boat brought us into small tributaries, where Scarlet and Blue-and-yellow Macaws (*Ara macao, A. ararauna*) flew across in amazing numbers. For me, the biggest avian thrill was a bird I had never seen before: the very peculiar Hoatzin (*Opisthocomus hoazin*) (plate 17). About the size of chickens, they perch above the water, uttering hoarse grunts and flying off so awkwardly that I imagine even *Archaeopteryx* might have been more graceful. I had known that this is a very special bird: the sole member of a family so different from other birds that its relationships were very uncertain. (It now is known to be a very ancient lineage, the sister group to almost all other land birds.[1]) I knew also that young Hoatzins escape perceived danger by dropping into the water and later climbing back up with the help of unique claws on their wings. And I knew that Hoatzins eat leaves.

What I didn't know then is that they are avian cows: they harbor bacteria in their large crop that break down plant cellulose into sugars that the bird can use, just as cows and other ruminant mammals do. Later in the 1990s, when DNA sequencing became easier, Allan Wilson's research group analyzed the enzyme lysozyme that the Hoatzin secretes in its stomach.[2] In most animals, lysozyme kills harmful bacteria by breaking down their cell walls. But in ruminant mammals a stomach lysozyme breaks down some of the useful digestive bacteria, providing the animals with a source of protein. It turned out that the Hoatzin does the same with its stomach lysozyme. The Wilson group discovered that the Hoatzin's lysozyme has several amino acid units that distinguish it from the enzyme in several other birds and mammals but which are identical to those in the cow. These amino acid differences enable the enzyme to function in the stomach environment. The Hoatzin, then, has evolved the same unusual biochemical adaptation as ruminant mammals, using much the same modifications of the same enzyme: convergent evolution at the biochemical level.

Birds have evolved countless features by natural selection, adaptations that enhance survival and reproduction in diverse ways and in diverse environments. The adaptive value of many structural features seems fairly obvious: long legs enable herons to forage in water, long bills enable ibises and sandpipers to extract small prey from mud or sand. Biochemical and physiological adaptations may be less obvious, but they have often been the key evolutionary changes that enabled birds to live in new environments and feed on new kinds of food. With DNA sequencing, the evolution of these features can now be studied in detail. I will describe a few such studies that I think are particularly interesting and then set them in a broader framework: what do we need to know to fully understand the evolution of birds' adaptations? How predictable is adaptive evolution? If life exists on any of the many planets elsewhere in the universe, is it likely that there is a parallel proliferation of birdlike creatures?

Collectively, birds eat almost everything they can overpower or swallow. Feeding adaptations of bills, tongues (grub-spearing tongues of woodpeckers, brushy tongues of nectar-feeding honeyeaters and lorikeets), and feet (notably in owls, hawks, and the Osprey with spiny toe scales) are legion. Biochemical adaptations are almost as diverse. For example, nectar-feeding hummingbirds can detect sugars, but how they do so turned out to be surprising.[3] We taste a sugar, such as sucrose, when a receptor protein in the membrane of taste cells changes shape as it binds the sucrose molecule. The cells (which are nerve cells) then send a signal to the brain. Most vertebrates have three genes (let's call them G_1, G_2, G_3) that are responsible for sweet and savory tastes. These three genes arose in the ancestor of vertebrates from a single original gene by a mutational process called gene duplication. Proteins encoded by G_1 and G_3 combine to form the savory receptor, while G_2 and G_3 proteins together form the sweet receptor. Most mammals have all three genes, but G_2 has been lost in some carnivorous mammals—and in birds, which have only G_1 and G_3. How, then, can hummingbirds taste sugar?

Maude Baldwin and her collaborators measured the responses of the taste receptors. They found that the G_1–G_3 receptor protein responds to amino acids, not sugars, in fishes, mammals (which was already known), and both chickens and Chimney Swifts (*Chaetura pelagica*). Swifts are hummingbirds' closest relatives, and there is little doubt that like them, the ancestors of hummingbirds fed almost entirely on insects. But the G_1–G_3 protein of hummingbirds responds to several sugars! This is a major change in the protein's function, which turns out to be based on multiple mutations in both the G_1 and G_3 genes that together changed the surface of the combined protein so that it reacts to sugars. The evolution of this protein is surely one reason for the great adaptive radiation of the hummingbirds, more than 330 species that feed largely on nectar.

If you have ever been on a "pelagic" trip in search of sea birds, you may have seen the leaders or crew "chum" with bits of fat or chopped fish, or simply by dripping fish oil onto the water, to attract birds. "Tubenoses" (Procellariiformes), such as storm-petrels and shearwaters, have a keen sense of smell and are especially responsive. These birds feed mostly on fatty seafood, and they digest complex fats much more efficiently than mammals do. Their several adaptations include high concentration of bile salts in the intestine, an ability to move food back up into the gizzard from the intestine so that it has a second round of digestion, and more efficient breakdown by enzymes. Similar adaptations enable Yellow-rumped Warblers (*Setophaga coronata*) and Tree Swallows (*Tachycineta bicolor*) to digest the wax coating of bayberries along the northeastern US coast during the winter and enable the Lesser Honeyguide (*Indicator minor*) to eat the wax of bees' honeycombs.[4]

Less exotic but still important physiological adaptations distinguish birds with subtly different diets. Thrushes eat lipid-rich (fatty or oily) fruits, while Cedar Waxwings eat mostly sugary fruits. Thrushes digest lipids more efficiently than waxwings do, and waxwings assimilate sugars much faster than thrushes do. At least some of their enzymes differ accordingly: waxwings express a sugar-digesting enzyme (sucrose), but American Robins (*Turdus migratorius*) and four other North American thrushes do not.[5]

Biochemical adaptations are also keys to living in some habitats. Two of the most awe-inspiring places I've visited were Lauca National Park in northern Chile, viewing Giant Coots (*Fulica giantea*) and three species of flamingos at

4,500 meters (14,760 feet) above sea level, and Balang Shan Pass in Sichuan, at 4,487 meters (14,722 feet). Here, my notes remind me, "I could walk only very slowly (even downhill was an effort), somewhat unsteady on my feet." In this spectacular Chinese landscape, we saw Rosy Pipits (*Anthus roseatus*), Alpine Accentors (*Prunella collaris*), and the beautiful Snow Pigeon (*Columba leuconota*), flying about as actively as its relatives in Manhattan.

How do birds, with their high metabolism and activity levels, manage at such low concentrations of oxygen? To be sure, I wasn't as fit as a mountaineer or even an average forty-year-old, but even a Sherpa[6] might be challenged to match the Bar-headed Goose (*Anser indicus*), which migrates between northern China and its wintering grounds in India by flying over the Himalayas, at 5,000 to 9,000 meters (16,404 to 29,527 feet) above sea level (plate 18). At these altitudes, the concentration (partial pressure) of oxygen is only about 30% of the sea level concentration. Compared with most other geese, the Bar-headed Goose has larger lungs, higher density of capillaries in heart muscle (which helps oxygen diffuse to the blood), and highly unusual hemoglobin, which can load oxygen in the lungs at very low concentration and still unload it in other tissues.

This species' oxygen management is achieved by three amino acid differences (compared with other geese) in the alpha chain of the hemoglobin molecule. At positions 18, 63, and 119 in the sequence, the Bar-headed Goose has amino acids G, A, and P, respectively, whereas all other species of geese have amino acids S, V, and A. (These are the standard abbreviations for the various amino acids.) The common ancestor of all geese must have had these shared amino acids, which changed in the evolution of the Bar-headed Goose. A research group led by Jay Storz, at the University of Nebraska, inserted the ancestral DNA sequence into bacteria, which then produced hemoglobin that the investigators could analyze. They then used a method called site-directed mutagenesis to change the ancestral DNA sequence, and therefore the encoded hemoglobin, altering the amino acids at each of these three positions, singly and in every combination. That is, they replayed every possible sequence of evolutionary change from S, V, A to G, A, P. At each step they tested the altered hemoglobin's affinity for oxygen.[7] (Very cool evolutionary chemistry!) The change at position 119 (from A to P) always increases O_2 affinity, whether it happens first or follows change at one or both of the other sites. The amino acid substitutions at positions 18 and 63 also increase O_2 affinity, but the size of the effect depends on whether the amino acid is the first to change or follows another. The important message is that the effect of a mutation can differ, depending on the company of other mutations: what happens in evolution can be contingent on what happened earlier. The process of adaptation may be harder if mutations have to happen in a certain order to be useful. (This has been shown in similar studies of other proteins, but not in birds.) The Storz team also discovered that the change from amino acid V to A at site 63 has a nasty "side effect." But the mutation from S to G at site 18 compensates, suppressing the harmful "side effect." Probably a lot of genetic changes are advantageous because they repair the harmful "side effects" of other mutations whose benefits came with costs.

What about other high-altitude birds? Most have evolved hemoglobins that, as in the Bar-headed Goose, have high oxygen affinity, but usually this is accomplished by different amino acid changes. In the Andean Goose (*Chloephaga melanoptera*), a single mutation in the beta chain of the hemoglobin acts the same way (in terms of physical chemistry) as one of the novel amino acids in the Bar-headed Goose's alpha chain. A study of eight species of high-altitude South American ducks found at least nineteen different amino acid changes that are thought to enhance O_2 affinity, with seven of these occurring independently in two or more species. Tits from the Tibetan Plateau (including the peculiar Ground Tit that I mentioned in chapter 2) also have a variety of hemoglobin differences from related lowland species, and a few of these mutational changes have occurred in parallel in two or more species. High- and low-elevation species of hummingbirds in the Andes tell the same story: the same biochemical pathways, such as cellular respiration, have been modified repeatedly in different lineages, but the specific proteins and amino acid sites often differ: there are some parallel, repeated evolutionary changes, but far more than one path of adaptation.[8] At the level of the whole, living organism, we can make broad predictions about evolution: in any group of birds, high-altitude species can function better at low oxygen levels than their lowland relatives would. But at the molecular level, evolution becomes far less predictable: there are many ways of achieving adaptation, even in a single critical protein.

That is a rather sweeping statement, a generalization about the evolution of adaptation—not just about life at high altitude, but about adaptation in general. It leads me to the second part of this chapter, in which birds illustrate the broad features of adaptation by natural selection.

———

If we were to understand fully the evolution of a certain characteristic, we would want to answer three kinds of questions. First, what has been its history? When did it evolve, and from what antecedent feature? Or was it truly new, rather than a modification of the old? Second, what was the biochemical or developmental process by which the feature has been changed? Look at the long tail feathers (rectrices) of a male Ring-necked Pheasant (*Phasianus colchicus*). Their great length results from genes that differ in their identity or activity from those that affect rectrix development in short-tailed birds such as quail. How do the genes produce this result? Third, we would like to know why the feature evolved: if by natural selection, what advantage did the feature provide? How did a long tail enhance survival or reproduction in the pheasant's ancestors?

Consider the quintessential feature of birds: the feather. Where did it come from, and by what steps has it evolved? The best current model, by Richard Prum and Alan Brush, combines evidence from fossils and feather development with a touch of speculation.[9] The word "feather" brings to mind a "pennaceous" feather with a stem, a shaft (rachis), and a vane on both sides, made up of many slender units (barbs) that bear small, complex projections fore and aft (barbules) that hook each barb to its neighbors. Pennaceous feathers include the flight feathers of the wings and tail as well as the contour feathers that cover the body. But birds have other kinds of feathers as well: for example, down feathers lack a rachis and have loose, soft barbules, so the barbs are separate and form a tuft. What they have in common, and what makes the feather uniquely different from mammalian hair or reptiles' scales, is that the stem and rachis are a hollow tube that develops from the periphery of a small bud (papilla) at the bottom of a pit in the skin.

The details of just how a feather develops are complex, but they led Prum and Brush to speculate that the first feathers were simple hollow filaments (figure 6.1). The next stage, they suggest, was like a down feather: a tuft with simple branches—barbs—that lacked barbules. Later still, barbules evolved, linking the barbs together to form the vane of a pennaceous feather. Under this hypothesis, feathers did not evolve for flight; instead, their original function and advantage may have been to provide insulation, or possibly other functions. At least this part of Prum and Brush's scenario is well supported by the many feathered dinosaurs that are now known. The oldest of these, such as *Sinosauropteryx*, have simple feathers that fit the postulated first stage; the later stages in the Prum–Brush scenario are found in later dinosaurs.

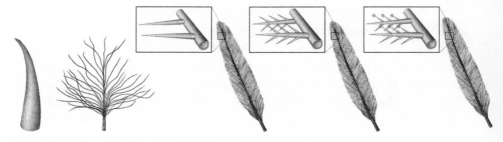

FIGURE 6.1. Several possible stages in the evolution of feathers. (From left) A simple hollow, hairlike structure may have been the first stage, carried into later feathers as the rachis. (Second from left) The formation of barbs (hollow branches off the rachis), perhaps with simple barbules, formed a structure much like the down feathers of modern birds. (Third, fourth, and fifth) Modifications of the barbules so that they hook each barb to its neighbors resulted in a coherent vane, as seen in most of the visible feathers of an adult bird. (After Prum and Brush 2002; art, Luci Betti-Nash.)

How did the very first simple feathers evolve? Some previous authors suggested that they were modified scales, and it is true that the earliest stage in the development of both scales and feathers is a "placode," a thick group of epidermal cells that forms above a concentration of dermal cells. But the subsequent development of scales is radically different: they form as small projecting folds of skin, unlike the tubular development from a sunken papilla. It is hard to see how the feather's mode of development could be a modification of the scale's mode. So the feather might have evolved as a truly new structure, engendered by a novel set of interactions of genes that determine development from the early placode.[10]

An evolutionary change in a structural feature is a change in the feature's development. Think about the webbed foot of a duck compared with the unwebbed, free toes of a chicken. As in other four-limbed vertebrates, the leg begins as a bud that develops into a paddle-like structure. Groups of cells condense in the paddle that eventually form cartilage and later, bone. These cell condensations are the embryonic digits. In the developing chicken foot, the cells in between the digits undergo "programmed cell death," resulting in the formation of free toes. Cell death does not occur in the developing duck foot: hence the webbing. The evolution of the webbed foot involved a change in the *expression of genes* that govern the process of cell death.

There are two kinds of evolutionary changes in genes' effects on organisms' features. The most familiar way, as we have seen, is by *coding changes* in the amino acids that make up the protein that a gene encodes. The other is a change in the *expression* of a gene in certain cells or parts of the body, or at certain times during development. All cells in an animal (or plant) have exactly the same set of genes as in the fertilized egg from which the cells stem. Most genes are "expressed," or activated, only in certain cells (e.g., visual pigment genes only in retina cells). "Expression" means that the gene is transcribed into messenger RNA (mRNA), a single-strand complement of one of the two DNA strands that provides the template for the sequence of amino acids in the encoded protein (recall figure 2.2). The level of gene activity can be measured in cells by measuring the mRNA level. Whether, and at what level, a gene is expressed is often regulated by another gene that produces a protein that binds to the target gene and turns that gene's transcription on or off. Sometimes that target gene's protein then regulates yet other genes; most structures develop as a result of a network of interacting genes that looks like a complex wiring diagram. Evolutionary changes in most structures seem to be based on changes in the regulation of sets of genes that determine just when, where, and for how long cells divide to build up a structure and what proteins are produced in those cells.

This background helps us understand how birds' wings evolved. The first land-living vertebrates, about 360 Ma (million years ago) had as many as eight digits, but their descendants soon settled on five, the number we share with the ancestors of crocodiles, dinosaurs, and birds. Their hand was much like ours, with several carpal (wrist) bones, five metacarpals (the long bones in the palm of our hand), and five digits (numbered I to V from front to back), each with two to four segments (phalanges, singular phalanx). Digits I and V were lost in the ancestors of theropod dinosaurs (see figure 3.4). Birds retain one phalanx in their "first" digit (actually II), which is movable and bears feathers (the alula). Some carpal bones and metacarpals III and IV are fused into a single unit, which articulates with digit III, with two phalanges, and the smaller digit IV, with only one. (Check it out if and when you eat a chicken wing.) Biologists have studied the development of limbs in chick embryos for almost a century and established that the development of digits, first as cartilage and then bone, is induced by a diffusing substance that emanates from the rear side of the limb bud. This is now known to be the protein encoded by the delightfully named *Sonic hedgehog* (*Shh*) gene, which up-regulates other genes, such as *Sox-9*, that control the development of cartilage. All these genes are expressed, and cartilage develops, in all five digit positions in the crocodile leg bud.[11] In birds *Sox9* is expressed and cartilage starts to develop in the digit V position but then stops. Digit I (which also fails to develop) is represented by *Sox9* expression in the chicken's wing bud but not in the African Collared Dove (*Streptopelia roseogrisea*) or Zebra Finch.[12] So bird embryos manifest the developmental beginnings of all five digits, and the skeleton of the modern bird wing results largely from changes in specific genes that terminate the development of certain digits and phalanges.

Just because a feature would be useful doesn't mean the right mutations will occur: the mutational process is blind to what would or wouldn't be advantageous. Because the development of many characteristics is complex, depending on many genes, there exist constraints that may prevent some ideal adaptations from evolving, or at least make them less likely. The most obvious cases are when a complex characteristic has been lost in evolution and can't re-evolve even if it would be advantageous. Diverse fish-eating vertebrates, such as dolphins, the slender-snouted crocodilian called the gavial, and extinct Cretaceous birds such as *Hesperornis* hold fish fast between rows of many teeth; however, all living birds are descended from an ancestor that lost teeth. No bird has ever re-evolved teeth—not even fish-eating mergansers, which have *ersatz* teeth: serrations along the edges of the horny bill sheath (figure 6.2).

In the same vein, there are many examples of "phylogenetically conservative" features that persist, seemingly unchangeable, throughout large clades over long periods of time. Certain lineages of antbirds (Thamnophilidae) live up to their name: they follow army ant swarms and feed on insects that attempt to escape the ants. This behavior evolved in the Miocene (23 to 6 Ma) and has not been reversed, or lost, by any species.[13] On a longer time scale, almost all birds have one of two toe arrangements: toe 1 behind and toes 2, 3, and 4 forward (the anisodactyl condition) or toes 1 and 4 behind, with 2 and 3 forward (zygodactyl). The huge order Passeriformes (songbirds), the Coraciiformes (kingfishers, hornbills, etc.), and some others all have anisodactyl feet without exception. The zygodactyl condition is uniform in several orders, including woodpeckers and their relatives (Piciformes), parrots (Psittaciformes), and cuckoos (Cuculiformes)—even in ground-dwelling species such as roadrunners. Perhaps remodeling would require going through a disadvantageous stage, like suspending elevator service for repairs. This would be impossible for natural selection, which can't look forward to a superior future outcome. (Because, of course, natural selection isn't an agent—it's just a difference in reproductive success, generation by generation.)

(A)

(B)

FIGURE 6.2. Complex structures, if lost in evolution, are generally not regained, but their function may be. (A) The Cretaceous marine bird *Hesperornis* had teeth that enabled it to grasp fish. (B) The bills of fish-eating mergansers have serrations that substitute for teeth. No living birds have teeth. ([A] From Everhart 2011, courtesy of Michael J. Everhart, Sternberg Museum of Natural History, Hays, Kansas; [B] from Futuyma 2013, courtesy of Oxford University Press / Sinauer Associates [original by Nancy Haver].)

We might suppose that there wouldn't be a disadvantageous intermediate stage between the two toe arrangements if a single mutation were to convert one into the other in a single big "jump," or "saltation." Whether or not evolution can happen by saltation is one of the most enduring questions in evolutionary biology. Most evolutionary biologists think that evolution almost always proceeds gradually by successive slight changes but acknowledge that there can be small jumps, or discrete changes (such as the single mutation from the pale gray ancestral peppered moth to the black form). Birds offer several lines of evidence that evolution is often gradual. First, variation within and between related species in structural features is often based on several genes, some of which contribute substantially and others only slightly to the overall difference. Multiple genes contribute to variation in bill size and shape in Darwin's ground finches (*Geospiza*) in the Galápagos Islands,[14] including differences in the expression of five genes known to affect skull and bill development.

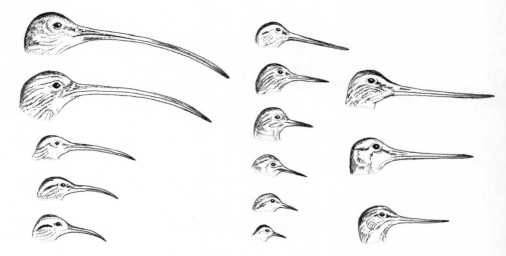

FIGURE 6.3. Graded differences in bill length of some members of the sandpiper family, drawn to scale. Lengths range from 18 millimeters in the Least Sandpiper (*Calidris minutilla*) to more than 200 mm in the Long-billed and Far Eastern Curlews (*Numenius americanus, N. madagascariensis*). (After Futuyma 2009, courtesy of Oxford University Press / Sinauer Associates. Art, Stephen Nash.)

Second, related species often display graded differences in a structure. For example, bill length in the sandpiper family ranges from the very short bills of stints to the disproportionately long bills of several species of curlews and godwits (figure 6.3; see also plate 12).

Third, distinctive higher taxa usually have acquired their features piecemeal, by "mosaic evolution." In the family Picidae (figure 6.4), the basal phylogenetic split is between the wryneck branch and the piculets and woodpeckers, which have a shock-absorbing hinge between the beak and the skull. The piculets lack the woodpeckers' stiff, strongly pointed tail feathers, enlarged terminal tail bone, and the ability to move the outer toe sideways, which helps woodpeckers to cling to tree trunks and climb upward. Among the woodpeckers, the basal split is between the genus *Hemicircus* (Heart-spotted [*H. canente*] and Grey-and-buff [*H. concretus*] Woodpeckers) and all others. *Hemicircus* species are intermediate: like piculets, they have a short, soft tail and cannot move the outer toe sideways, but they have an enlarged tail bone and most of the other features of typical woodpeckers.[15]

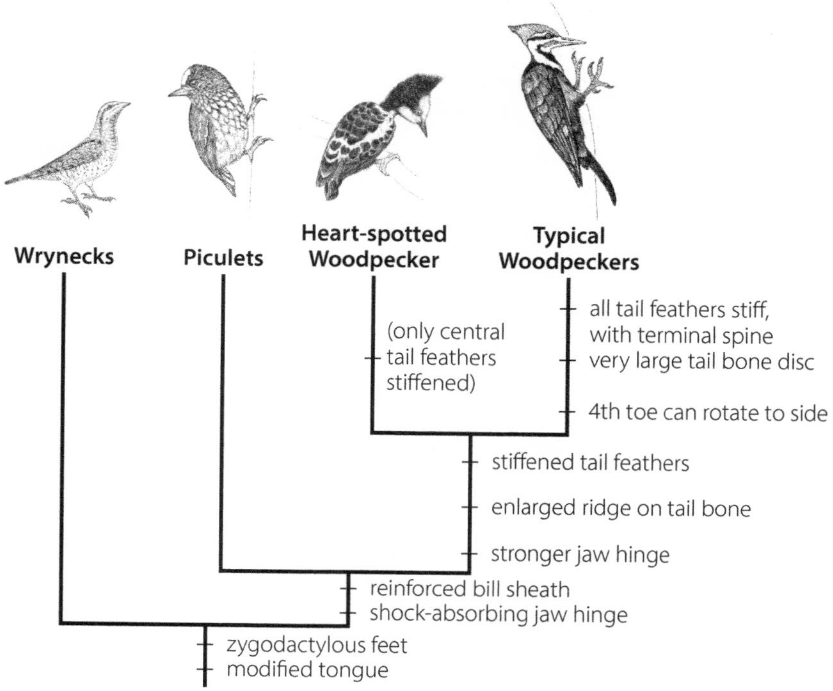

FIGURE 6.4. Stepwise, mosaic evolution of adaptations for climbing trees and drilling in wood in the woodpecker family. From left to right: Eurasian Wryneck (*Jynx torquilla*), Ocellated Piculet (*Picumnus dorbignyanus*), Heart-spotted Woodpecker (*Hemicircus canente*), Pileated Woodpecker (*Dryocopus pileatus*). Notes along each ancestral branch mark the evolution of features shared by descendants. Wrynecks lack the adaptations woodpeckers have for climbing and drilling in wood. Piculets have some skeletal adaptations for wood drilling. The Heart-spotted Woodpecker is intermediate in structure between piculets and typical woodpeckers, with their modified tail feathers, enlarged tail bone (pygostyle), and ability to move the fourth toe to the side. (Art, Stephen Nash.)

Fourth, species with a feature that differs extremely from any other bird are usually very distantly related to anything else. During their millions of years of separate evolution, any intermediate species that might have existed have become extinct instead of forming persistent branches that can show the intermediate steps (the way piculets and *Hemicircus* woodpeckers do). For example, flamingos, with their highly specialized bill and tongue, separated from their nearest living relatives, the grebes, almost 40 Ma (see figure 3.6). The very distinctive characteristics of both groups probably evolved by stages that haven't been retained by any living species.

The multitudinous adaptations of birds match the variety of their ways of life: their ecological niches. Birds are specialized for different habitats, diets, and ways of locomotion, foraging, migration, and more. A specialized species is sometimes subjected to natural selection that can lead to evolutionary extremes. For example, species of hummingbirds vary greatly in the length and curvature of their bill, and the differences are at least broadly related to the length and shape of the flowers they most commonly use.[16] Possibly the most extraordinary species—one of the most memorable birds I've ever seen—is the Sword-billed Hummingbird (*Ensifera ensifera*) in the Andes (plate 19). Its bill is up to 11 centimeters long, often longer than its body. It feeds on various flowers, especially in the passionflower family, with equally long nectar tubes.[17] How and why did plant and bird become so extreme? Through a coevolutionary process that Darwin postulated in 1862, long before there was any evidence. Darwin realized that although plants provide nectar for their pollinators and the pollinators help plants to reproduce, this mutual aid is no more altruistic than my relationship to a shopkeeper. We both give only because we need to receive. A bird whose bill is longer than the flower tube has the advantage that it can be sure to obtain all the nectar at the bottom. But effective pollination occurs only if the flower tube is at least as long as the bird's bill, if not longer, because then the bird is forced to push its face into the flower and contact the stamens and stigma. So there is natural selection for slightly longer flowers. But then birds with slightly longer bills have an advantage—and so on, in a continuing transactional conflict, with flower and bill lengths escalating like mobile phone costs. No one has tested this idea with hummingbirds, but Anton Pauw and his collaborators found strong evidence for Darwin's scenario by studying a fly species with an extremely long tongue that visits equally long flowers. Flies with longer tongues consume more nectar, and longer flowers (in a species of iris) receive more pollen. A similar study of a long-tongued bat provided similar evidence.[18]

The bill differences among species of hummingbirds, sandpipers, and Darwin's finches are adaptations to their different specializations: their different diets and ways of foraging: Why do species become specialized? Why not do it all?[19]

A common reason, possibly the most common, is that "a jack of all trades is master of none." There are commonly trade-offs between different tasks. As far as I know, no champion swimmer is also a champion weight lifter. Often, the features that best adapt a bird for one activity reduce its proficiency in other activities. We have already seen that the length and shape of wings differs among species with different modes of flight, and that in Darwin's finches, crossbills, and flowerpiercers, different foods are best processed by bills that differ in size and shape.

Trade-offs may explain why species specialize, but why do they evolve different specializations? Even the most similar species often prove to have somewhat different ecological habits (or ecological niches). In one of the most famous studies of niche differences, Robert MacArthur, who had a huge impact on ecology, studied five species of warblers, in the same genus, that are similar in size and shape and coexist in spruce and fir forests in northeastern North America. He found that they differ in feeding behavior (for example, the Cape May Warbler [Setophaga tigrina] sallies for flying insects more than the others, while the Black-throated Green [S. virens] rummages through dense foliage) and in the parts of the tree that they most use. The Cape May forages high in the tree, on the outside; the Blackburnian (S. fusca) is similar but extends its foraging to lower levels; the Yellow-rumped (S. coronata) is generally inside the tree at lower levels, even down to the ground.[20] Probably these species are obtaining somewhat different sets of insect species, although MacArthur did not pursue that.

One of the most important reasons that related species diverge in this way is competition for food or other resources. In On the Origin of Species, Darwin wrote that "competition will generally be most severe between those forms which are most nearly related to each other in habits, constitution, and structure," so that "the more diversified these descendants [of an ancestral species] become, the better will be their chance of succeeding in the battle for life." Those variant individuals of a species that compete less with other species will tend to have higher fitness, so their habits and traits become more prevalent, and the species will evolve to differ more from its competitors.

There are two major kinds of competition. "Contest" competition involves outright aggression between individuals. This has been documented for at least 270 pairs of bird species, and the pattern is consistent: the larger one almost always wins.[21] Competition matters: subordinate species tend to nest later than related dominant species, to have higher mortality rates, to lay larger eggs, and to be more specialized in their foraging behavior. Scott Robinson and John Terborgh found that among Amazonian birds, aggression occurs in almost all the pairs of closely related species that have nonoverlapping territories; the Black-throated (Myrmophylax atrothorax) and Chestnut-tailed

Antbirds (*Sciaphylax castanea*) are an example. But species whose territories overlap usually use different microhabitats and resources, and these species aren't aggressive toward each other—as in the Sooty Antbird (*Hafferia fortis*), which feeds at army ant swarms; Goeldi's Antbird (*Akletos goeldii*), which lives in vine tangles; and Plumbeous Antbird (*Myrmelastes hyperythrus*), which lives in stands of a banana-like plant, *Heliconia* (plate 20). Might they have evolved different habits to reduce competition? That is what Darwin would have expected.

"Scramble" competition can occur simply because each species reduces the supply of food or nesting sites: the early bird gets the worm. This is why Darwin supposed that closely related species would adapt to different resources. There is plenty of evidence that species of birds compete.[22] Populations of Great Tits and Blue Tits have been experimentally manipulated by supplying extra nest boxes (allowing increased density of birds) or removing nests from boxes (causing decreased density). As a Blue Tit population in England grew from one year to the next in one such experiment, Great Tits declined; in Hungary a similar experiment showed that Blue Tits cause Great Tits to produce lighter nestlings. In Sweden these species together reduced the survival of Collared Flycatcher offspring. In some cases, the impact of nest predators on one species is increased by the presence of another. This was one reason why Virginia's Warblers (*Leiothlypis virginiae*) were more successful in breeding when Orange-crowned Warblers (*L. celata*) were experimentally removed.

Some of the most important evidence for Darwin's suggestion that natural selection would cause species to diverge in their use of resources has come from birds that show ecological character displacement.[23] This refers to a pattern in which populations of two species that are *sympatric*—in the same geographic region—differ more in their use of resources than separated (*allopatric*) populations of the same two species.[24] Often the sympatric populations differ more in bill or body size, a likely indication that they differ more in diet than do the allopatric populations. Jon Fjeldså studied specimens of all species of grebes that coexist locally with a closely related species. In several cases, he found that a population that is sympatric with another species differs from allopatric populations.[25] For example, the fish-eating Silvery Grebe (*Podiceps occipitalis*) has a broad distribution in Andean lakes and co-occurs with the Junin Grebe (*P. taczanowskii*), which is restricted to Junín Lake in Peru. The Junin Grebe has a long bill and feeds on small fish. The Junín population of the Silvery Grebe has a smaller bill than other Andean populations, and instead of fish, it

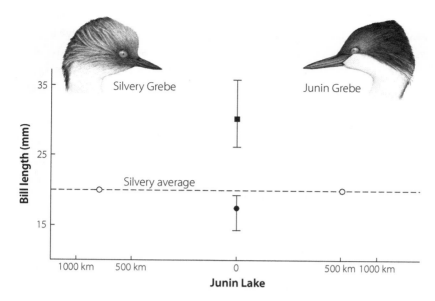

FIGURE 6.5. Character displacement in grebes. The bill lengths of two species were measured in samples at various distances from Lake Junín, where the long bills of Junín Grebes (*Podiceps taczanowskii*) are shown by a square (average length) with "whiskers' showing variation. The horizontal broken shows the average bill length of Silvery Grebes (*P. occipitalis*) along the transect. The average bill length of the population in Lake Junín is lower, indicated by the filled circle. Here, the Silvery Grebe has evolved to become more different from the Junín Grebe. (After Fjeldså 1983; art, Luci Betti-Nash.)

eats zooplankton and midges picked from the water surface (figure 6.5). It appears to have "evolved away" from the Junin Grebe.

Of course, competition with other species isn't the only reason a species may adapt to and become specialized for a certain food or habitat: there will be an advantage in using any abundant resource that isn't already monopolized by other species. Many Australian forests have an abundance of mistletoe, so it isn't surprising that its fruit is the main food of the Mistletoebird (*Dicaeum hirundinaceum*) and the Painted Honeyeater (*Grantiella picta*). But how does the shift to a new resource happen? What are the first steps?

During a summer drive along the shore of New York's Long Island some years ago, I noticed Laughing Gulls (*Leucophaeus atricilla*) milling around in the air like giant swallows. They were catching winged ants—not a habit I had ever associated with gulls. They embodied a hypothesis that was best stated by the great ornithologist and evolutionary biologist Ernst Mayr, who wrote that "the evolutionary changes that result from adaptive shifts are often initiated by a change in behavior, to be followed secondarily by a change in structure. It is very often the new habit which sets up the selection pressure." Allan Wilson, the early leader in molecular evolution whom we have already met, suggested that birds' capacity for behavioral innovation and social propagation of new habits was the driving force in the rapid evolution of their great anatomical diversity.[26] Probably the most remarkable innovations are tool-using by the Woodpecker Finch (*Camarhynchus pallidus*), which probes for insects with a spine or short twig, and the astonishing New Caledonian Crow (*Corvus moneduloides*), famous for making tools for the same purpose and for solving complicated tasks in the laboratory.[27] Louis Lefebvre's research group at McGill University compiled the many published reports of innovative feeding behaviors, like my swallow-mimicking gulls, and found that the frequency of innovation in various groups of birds is correlated with the size of certain parts of the brain—and with the number of species. Species for which innovative behaviors have been described also are more successful when introduced into new regions or habitats, and they have a lower risk of extinction.[28]

In many species, individuals seem to learn to be proficient at one feeding behavior or another and develop a specialization. Among the group of Darwin's finches, one species lives on the remote Cocos Island, with only three other land bird species. The Cocos Finch (*Pinaroloxias inornata*) has been "ecologically released," playing ecological roles that on a continent would be filled by diverse species. Tracey Werner and Thomas Sherry watched color-ringed birds for ten months and found that each bird forages mostly in only two or three out of nine foraging modes. For example, one bird devoted 87% of its effort to gleaning leaves for insects, another 58% to probing clusters of dead leaves, and a third 42% to extracting nectar from plants' nectar glands.[29] Such learned innovations sometimes spread through bird populations, much as they do in humans, by social learning; in chapter 4, I described how the learned habit of getting cream from milk bottles spread through populations of Blue and Great Tits. It is easy to imagine that a whole population could learn to use a new resource. Natural selection would then favor structural or other features that increase their efficiency, leading to divergent evolution. Maybe someday there will be sparrow populations specially adapted for clinging to bird feeders.

You may have thought of other features of birds that call for explanation. Can evolution by natural selection explain why some species lay many eggs but albatrosses only one? Why birds of paradise are so spectacularly and diversely adorned? Why females are more brightly colored than males in phalaropes and some other groups? Why birds flock together, sometimes with many other species? In the next three chapters, I consider the diverse life histories, sex lives, and social relations that add to our endless fascination with birds.

7

Owls and Albatrosses

LIFE CYCLE EVENTS AND VARIATIONS

In midafternoon on December 29, 1999, I sat down behind a shrub in Hattah-Kulkyne National Park in northwestern Victoria, Australia, and waited. I faced a circular mound, depressed in the middle, about 1 meter high and 4 meters across. After about forty-five minutes, the bird suddenly appeared: a male Malleefowl (*Leipoa ocellata*), about the size of a pheasant, who faced outward from the rim of the mound and began vigorously kicking debris into the depression. He slowly moved clockwise around the rim, kicking away, until, having traversed about 120°, he caught sight of me and ran off (plate 21).

Thirteen years later, I joined a quest for another intriguing member of the megapode family on the Indonesian island Sulawesi. At about nine in the evening, our birding tour group was ferried in small outrigger canoes across a magically moonlit mangrove-lined lagoon to an offshore beach. After a short walk our local guide spied and spotlighted a female Moluccan Megapode (*Eulipoa wallacei*), who had flown down from the looming mountainside to dig a pit in the sand, lay her eggs, and leave, never to see her offspring except by chance.

The reproductive behavior of megapodes, a family of Australasian galliform birds (chicken relatives), is strikingly different from that of any other bird. The Moluccan Megapode's eggs develop in solar-heated sand; a male Malleefowl, with some help from the female, builds a great mound—a compost pile—in which the eggs are warmed by a combination of insolation and heat emitted from decaying vegetation. Then the male alone tends the mound twice daily, changing the depth of the material in order to maintain a fairly constant temperature near 30°C. Megapodes' eggs are large, with abundant yolk that supports development to an advanced stage: the emerging chick, unlike all other new-hatched birds, has fully developed wings and, after digging its own way to the surface, walks or flies off on its own. A Moluccan Megapode chick loses no time in flying from its natal beach to the mountain slope, hundreds of meters away.

Species of birds vary greatly in whether one or both sexes, or neither, provide parental care. In most species, both parents care for the young, but only females feed and protect the brood in some groups (such as grouse and birds of paradise) and only males provide care in others (including ostriches and some shorebirds, such as phalaropes). Other than digging a pit and covering the eggs with sand, the Moluccan Megapode provides no care to its offspring. The burden of caring has been completely shed by brood parasites, such as cowbirds, honeyguides, and many cuckoos. Birds also differ in "life history traits": how long they live, at what age they start to reproduce, and how many eggs they lay. To give a few examples: Japanese Quail (*Coturnix japonica*) live for about six years in captivity and Zebra Finches twelve, while Sulphur-crested Cockatoos (*Cacatua galerita*) have reached at least fifty-seven years and Andean Condors (*Vultur gryphus*) seventy-nine.[1] The number of eggs in a clutch is highest in some gallinaceous birds (more than twenty in Grey Partridge [*Perdix perdix*]), while a Wandering Albatross (*Diomedea exulans*) lays a single egg every other year. Most small birds start reproducing within their first year, but gulls may take two, three, or four years to mature, and Wandering Albatrosses usually start when they are nine or ten years old.[2]

Evolutionary biologists would like to understand why various life histories—often called "life history strategies" without any imputation of awareness—have evolved.[3] What are the conditions under which these various life histories would be advantageous and likely to evolve by natural selection?

We might ask what accounts for a species' average number of eggs (clutch size). We know that females that lose one or two eggs to a predator often can replace them, so they are able to lay larger clutches. Why don't they?

Before the 1950s some biologists would answer that reproduction evolves to offset mortality, so that the species will survive. If mortality is high, more offspring are needed to maintain the population. But a British ornithologist, Oxford professor David Lack, pointed out that this hypothesis is completely at odds with Darwin's theory of natural selection and its modern genetic version. Darwin argued that if some *individuals* have a variant characteristic that increases their capacity for survival or reproduction, and if that feature is inherited, the characteristic will inevitably spread through the species population as generations pass and will become a characteristic of the species. In today's version of Darwin's theory, a gene mutation (allele) that alters some feature of an organism will increase in frequency, and generally replace the previous "standard" allele, if individuals with that allele have higher fitness. In chapter 4, I described how the fitness of a genotype equals its probability of survival, multiplied by its number of offspring if it does survive. A mutation that increases the number of offspring will increase (all else being equal), and the species will evolve a greater clutch size. The prospect of future extinction of the population plays no role—and can't. The fate of the population is a by-product of natural selection among genetically different individuals.

David Lack initiated the study of life history evolution in a 1947 paper in the ornithology journal *Ibis*.[4] He understood that natural selection would favor ever larger clutches, all else being equal. He proposed that not all else is equal. Specifically, a genotype that lays too many eggs would produce fewer, not more, surviving offspring if a limited amount of food were divided among so many nestlings that some or all would be weak and unlikely to survive. He studied European Starlings (*Sturnus vulgaris*) and Great Tits and reported some evidence of higher mortality of nestlings in nests with more eggs. But natural variation has so many simultaneous causes that such data can be very messy and inconclusive. An experiment—manipulating or treating some individuals but not others—was needed. Lack's student Christopher Perrins increased clutch sizes in nests of Common Swifts (*Apus apus*), which usually lay two or three eggs, by transferring single hatchlings from one nest to another, creating four-nestling broods. He repeated this for four years, and in every year he found, as Lack had predicted, that more surviving offspring fledged from three-nestling broods than from nests with four (figure 7.1).

Probably every benefit comes with some kind of cost, or trade-off. Lack proposed, and Perrins documented, a trade-off between number of offspring and offspring survival, presumably because the parents could supply only a limited amount of food. Later experiments by other researchers with other species of birds showed that the trade-off is often between clutch size and other aspects of fitness; in some cases, the parents suffered higher mortality or were unable to rear a second brood if their first brood was too large.[5] When one or two eggs were added to nests of Collared Flycatchers in Sweden, the fledglings weighed less and were less likely to return the next year as new breeders. Those that did return laid fewer eggs: apparently the stress of inadequate food when young carried over into adult life. What's more, females who

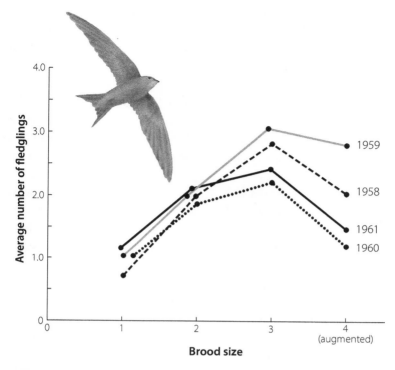

FIGURE 7.1. The number of fledglings from nests of Common Swifts (*Apus apus*) with different numbers of eggs. The natural clutch size is one to three; the investigator added eggs to make four-egg clutches. The experiment was repeated in four years, and in each year, natural three-egg clutches yielded more surviving fledglings than four-egg nests. (After Perrins 1964; art, Luci Betti-Nash.)

had to rear an augmented number of young lost weight and laid fewer eggs the following year.[6]

The evidence, then, supports the proposal that birds lay as many eggs as they can, subject to the constraint that too many will lower, rather than increase, the parents' fitness because of trade-offs: reduced survival of the offspring themselves or of the parents' own survival and subsequent reproduction. Now the question arises: what accounts for the variation in clutch size among species?

This is a matter of the species' ecology, so there is no simple answer. At one extreme are albatrosses, especially the very large Wandering and Royal Albatrosses (*Diomedea* species) They do not start reproducing until the age of nine or ten (six at the earliest); they lay one large egg, and that reproductive episode, from nest-building to the offspring's fledging and departure, takes almost a year (plate 22). Because food is sparse in large tracts of ocean, the parents fly as much as 15,000 kilometers in trips that last up to thirty-three days, returning with a crop full of food. The investment of energy in rearing just one chick is enormous, and it takes a year to recover before starting the process again.[7] Recovery includes replacing worn flight feathers, which takes so much energy that there is a trade-off between breeding and replacing the feathers on which their lives depend.[8] Instead of trying to rear more than one offspring, it is advantageous to produce one offspring every two years throughout the birds' usually lengthy life (often twenty-seven years or more).[9] Albatrosses have a "slow" life history: they are long-lived, start reproducing only after several years, and have few (one) young at a time. Most songbirds, in contrast, are "fast": they start reproducing when they are young (often within their first year), lay larger clutches, and tend to have shorter life spans. To fully make sense of these patterns, we need to understand the evolution of life spans.

It's an oversimplification, but let's distinguish intrinsic from extrinsic causes of death. As we humans age, we undergo senescence—various forms of functional degradation—and the chance of death increases with age, owing to cancer, heart attack, or many other systemic failures. These are "intrinsic" causes of mortality. On the other hand, car collisions, tsunamis, and mass shootings are "extrinsic" causes of death, in which the chance of death is fairly independent of one's age. For birds, inclement weather during migration, predation,[10] and window strikes are extrinsic mortality factors that operate at a given rate in any particular environment and can act at any age. Suppose the environment is harsh, and the chance of being killed is 50% every year. Then for every 1,000 birds in year 1, 500 will survive to year 2, 250 to year 3, and so on. At that rate, only about two birds will survive for ten years. In a more favorable environment, the annual chance of death from extrinsic causes might be 10%, and about 387 out of 1,000 initial birds would live for ten years.

The level of extrinsic mortality—how dangerous a bird's world is—affects the evolution of the species' potential life span: how long it lives (if free from those extrinsic dangers) before senescent physiological failure sets in. Most of the data on birds' "intrinsic life span" is based on captive birds, protected from the vicissitudes of the natural environment. That's how we know that cockatoos can live longer than quail and Zebra Finches. Why has the Zebra Finch evolved a potential life span of only twelve years?

Suppose the extrinsic mortality rate is high and there is only a very slight chance of living for twelve years. Then, mutations that would prolong the potential life span to fifteen or twenty would have no selective advantage, any more than you would gain from a savings plan that paid out a handsome sum at age 110: the chance you would live to use it is negligible. So species that live in harsher, more variable environments can evolve a short life span because a longer potential life is seldom realized. Now combine this with the "cost of reproduction" that Lack first postulated: investing a lot of energy in reproduction at one time can reduce the chance of survival and subsequent reproduction. In a very harsh environment, or one where the death rate from predation is very high, a high reproductive effort may reduce a bird's potential life span, but the bird's chances of living long are rather low anyway.[11] So we would expect that species that evolve in harsh environments would have a "fast" life history: reproduce early in life and have many offspring but undergo relatively early senescence and live relatively short lives. In contrast, species in more benign or constant environments, if they experience low rates of extrinsic mortality, would evolve a "slow" life history: a longer life span with reproduction continuing at advanced age—but a lower number of offspring per clutch because a larger number would reduce survival.

Like humans, birds undergo senescent decline, and it is accelerated by the effort of reproducing. Thomas Reed and colleagues studied a colony of Common Murres (Common Guillemots, *Uria aalge*) at the Firth of Forth in Scotland for twenty-two years, following the histories of banded, individually marked birds.[12] The life span during which birds reproduced averaged fifteen years. The birds underwent senescent decline in the three years before they died: they had lower breeding success and were less likely to defend a cliff-ledge nest site successfully against competitors (figure 7.2). The higher their reproductive success earlier in life, the faster they underwent senescence. The yearly chance of death increases with age in Western Jackdaws (*Coloeus monedula*) and was accelerated when investigators increased brood size by adding eggs.[13] So there is evidence for the theory of a trade-off between reproductive effort and intrinsic life span. Now let's return to the evolution of clutch size.

In the 1940s, Reginald Moreau, studying birds in Africa, and Alexander Skutch, in Costa Rica, both discovered that tropical songbirds generally have fewer eggs per nest than their temperate-zone relatives.[14] Many later studies confirmed this pattern. Tropical Southern House Wrens (*Troglodytes aedon musculus*), for instance, lay smaller clutches of eggs than Northern House Wrens (*Troglodytes aedon aedon*).[15] This discovery, of course, raised the question, why? Two possible answers have been thought most likely. One

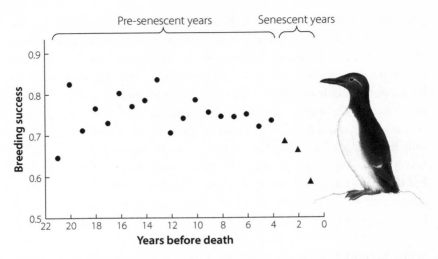

FIGURE 7.2. Senescence in Common Murres (*Uria aalge*). Each point is the average breeding success (percentage of eggs that survived to fledging) of 238 murres that were studied over their lifetimes. It dropped rapidly in the last three years of life. (After Reed et al. 2008; art, Luci Betti-Nash.)

implicates nest predation, which is the major cause of reproductive failure in songbirds worldwide.[16] The more nestlings a nest has, the more often parents must come to feed them, and the more likely they are to draw the attention of nest predators, so nest predation can create natural selection for smaller clutches. The rate of nest predation is generally higher in tropical than higher-latitude species, so this may help to explain the lower clutch size of tropical species.

The other hypothesis relates clutch size to adult longevity: tropical species tend to live longer.[17] The slow–fast contrast in life histories tends to distinguish tropical from temperate-zone species of passerine birds. So this idea is that adult birds in the tropics suffer less extrinsic mortality, have therefore been able to evolve longer potential life span, and can achieve higher fitness by repeated reproduction. But they must invest less effort each time they reproduce so that they don't increase their own chance of death. Thomas Martin, at the University of Montana, has found evidence for both of these hypotheses in data from multiple songbird species in Arizona, Venezuela, and Malaysia (figure 7.3).[18] Martin concluded that it has been advantageous for tropical species to evolve a long life span, with repeated reproduction, and that they achieve this by limiting their parental effort to a few nestlings that are so well

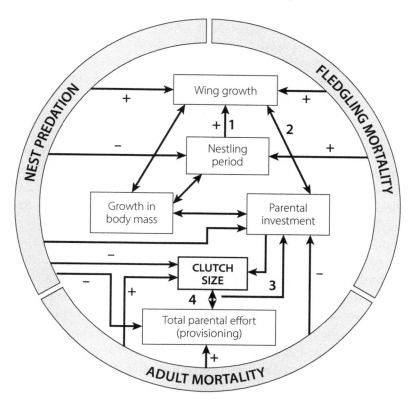

FIGURE 7.3. Martin's model of evolution of clutch size in tropical versus temperate-zone passerines, with arrows indicating the effect of one variable on another. Plus and minus signs indicate positive or negative influences. Fledgling mortality due to predators favors a longer nestling period that allows greater growth of wings (1), but longer nestling period is opposed by nest predation. Increased parental investment compensates to increase growth of wings (2), but greater parental investment requires greater total parental effort (3), which can expose parents to greater adult mortality. Reducing clutch size will reduce total parental effort (4), enabling greater parental investment in fewer offspring, so enhancing their growth of wings and ability to survive. (After Martin 2015, reprinted with permission of AAAS.)

cared for that they are more likely to survive the risk of predation both before and after they leave the nest.

Food delivery is important for the evolution of clutch size because the supply is limited—and so, nestlings compete for food. This competition has some intriguing—and maybe disturbing—evolutionary consequences. One effect that won't surprise anyone is that it can affect the evolution of the nestlings' appearance.

Among evolutionary biologists, Mary Jane West-Eberhard, a friend since graduate school, is one of the most original thinkers I know. She is an expert on the evolution of social insects, but her interests expand in many directions, including the evolution of social signals. In 1983 she noted that in many birds and other animals, "offspring compete strenuously for parental attention," and that because parents are often "in a position to exercise favoritism, or parental choice, in treatment of their offspring," they create what she called *social selection* for the "highly specialized and exaggerated signals" that the offspring often employ.[19]

West-Eberhard's suggestion was fully supported in its first experimental test, by Bruce Lyon, who studied American Coots (*Fulica amaericana*).[20] The chicks of this species have blackish plumage, from which long, orange filaments protrude on the head, neck, and upper back (plate 23). A pair has about six to twelve chicks, and Lyon found that a third to a half commonly die of starvation. Lyon speculated that the orange plumes evolved because they attract parents' attention when the chicks are competing for food items. He trimmed the orange plumes of half the chicks in twenty-one broods and found that, sure enough, the trimmed chicks, now dull black, were fed less often than their gaudier siblings, grew more slowly, and had a lower rate of survival.

Lyon suggested several possible reasons that adult coots might respond to the orange plumes, but as far as I know, this is still an unanswered question. One possibility is simply that the brain is programmed to register more intense stimuli, of any kind. In the 1950s Niko Tinbergen, a founder of the study of animal behavior, described abnormally intense stimuli that elicit a strong response. He called them supernormal stimuli. In a famous study, Tinbergen showed that Herring Gull (*Larus argentatus*) chicks peck at the red spot on their parent's bill to induce the parent to regurgitate food—but they peck more vigorously at a dummy with three red spots than one. Nestlings of most passerine species have yellow or orange mouth linings. When investigators dyed the mouth lining of Great Tit chicks red, the parents fed them twice as much food as normal chicks, which have orange-yellow mouths.[21] Apparently these birds have an intrinsic response to red that evolved for some other, as yet unknown reason, perhaps in an ancient ancestor of modern passerines. Some psychologists propose that humans respond to supernormal sexual and other stimuli, a propensity that is unremittingly exploited in advertising.

One way of competing with your nest mates is to attract your parents' attention. How else might you get an edge over your siblings? You could kill them (plate 24). Remember that a mutation will increase in frequency and become fixed in a species if its effect on some trait causes higher survival or reproduction than the prevailing genotype. Murderous behavior in a chick can be just such a trait. Siblicide is not only conceivable, it is common in many raptors, boobies, and herons and egrets and occurs in a variety of other species, such as the Black-billed Magpie (*Pica hudsonia*).[22] And it is not only frequent but "obligate"—it always happens—in some eagles, the American White Pelican (*Pelecanus erythrorhynchus*), and the Whooping Crane (*Grus americana*). (This was convenient for the recovery program, in which otherwise doomed second eggs of this highly endangered species were removed, to be reared by captive Sandhill Cranes [*Grus canadensis*] as foster parents.) In these species the parents do not intervene; they do nothing to stop the murder happening in full sight. Why should they? They cannot rear two vigorous offspring; their fitness will be higher if one kills the other and is healthy enough to survive fledging and the stresses of independent life. Siblicide and parental "approval," so to speak, are surely adaptive and have evolved by natural selection. (This is just one of many examples showing that we shouldn't even think of looking to nature for ethical or moral guidance.)

In some species the production of offspring appears almost designed to insure competition and death of some of them. You may have seen photos of broods of nestling Barn Owls (*Tyto alba*): the largest are often nearly twice as large as the smallest and are far more likely to survive. The disparity arises from the female's pattern of incubation. Most birds do not start incubating their eggs until they have laid an almost full clutch, so most of the embryos effectively start developing at the same time. But a female Barn Owl lays up to fourteen eggs, one every two or three days, and incubates them as soon as the first is laid.[23] Naturally the early offspring have a developmental head start over the later ones; they hatch earlier, are fed, become quite large by the time the last egg hatches, and have a much greater chance of surviving.

Female Barn Owls seem to "overproduce" eggs, laying more than are likely to be successful. So do female American White Pelicans, egrets, and all the other species that typically lose a chick to siblicide. The most extraordinary case is in the Rockhopper Penguin (*Eudyptes chrysocome*) and other crested penguins.[24] They lay two eggs, the second larger than the first, and they start incubating only when the second egg is laid. The chick from the second egg is much larger from the start and gets more food from the parents simply by reaching higher toward the parent's bill. It is usually the only survivor.

Why do females overproduce eggs? There are three widely accepted hypotheses. I suspect the least frequent advantage of overproduction is that the "extra" egg or nestling might be useful to its siblings or parents. As a possibly extreme example, nestling American Kestrels (*Falco sparverius*) fairly often cannibalize one of their number, especially when the parents deliver less food.[25]

The second hypothesis is "insurance," which is most obviously relevant to species that lay only two or a few eggs. The parents may be able to rear only one offspring, but if they laid a single egg, and the egg were destroyed or the nestling died, they would have zero reproductive success that year. In many such species, the smaller second offspring is almost certainly doomed—unless the firstborn (I mean first laid) has an accident or is frail for some reason. This idea has solid support. For example, the second egg of the Red-footed Booby develops into a surviving fledgling after the first egg's failure in 19% of clutches.[26]

The advantage of overproduction that I suspect has been important in more species is that there are good years and bad. It is hard to predict whether rodents will be abundant or sparse. If abundant, a Barn Owl pair produces a lot of healthy offspring that will carry on their genes; if sparse, at least a few will survive, and the others are just unlucky. This is a common pattern among owls, at least at high latitudes. Birders in the north temperate zone look forward to the occasional winter when Snowy Owls (*Bubo scandiacus*) move south in abundance: years when lemmings are common in the Arctic and a high percentage of owlets has survived.[27]

This leads me into another topic. It's fascinating, but a little complicated. I suggest a break—maybe to see or hear some birds.

———

The theme of offspring competition for parental care raises the question, who provides the care: mother, father, or both? More than 90% of bird species have biparental care: both sexes contribute care in one way or another (although in many, the female invests more effort than the male). Species with female-only care are known in at least eleven families. These include some groups with precocial young, such as pheasants and ducks, in which offspring forage soon after hatching and require mostly protection, rather than feeding, by their parents. Several families with female-only care, such as hummingbirds, cotingas, manakins, birds-of-paradise, and bowerbirds, are renowned for the males' elaborate plumages and displays. In contrast, the male incubates the eggs and guards the young in the ratites (e.g., ostrich, cassowaries, kiwis) and provides most or all of the care in many shorebirds, such as Eurasian Dotterel (*Charadrius morinellus*), Spotted Sandpiper (*Actitis macularia*), phalaropes, jacanas, and painted snipes. In these shorebirds and some other species with male care, females often court and compete for males and are often more brightly colored. How do we account for the evolution of these different family arrangements?[28]

Parents often cooperate in rearing their young—to a point. They also are in conflict, because caring for offspring has both benefits (passing on your genes) and costs. We have seen that caring for young takes a toll on parents, often reducing their chances of survival and subsequent reproduction. It also creates an "opportunity cost": staying at home with the kids reduces the opportunity to find more mates and have more offspring (that carry your genes). So each parent potentially gains by leaving as much work as possible to the other parent (figure 7.4): the greater the caregiving effort provided by one parent, the lower the effort of the other parent need be, if it still yields the greatest number of surviving fledglings. If offspring survival requires more care than one parent can dependably provide, natural selection in both sexes will favor caregiving to enhance both parents' fitness. Does this theory explain why species differ in whether females, males, or both care for their offspring? We have to know what influences the cost and benefit of caregiving versus defecting—leaving your mate to care for the young.

Because caring for the offspring certainly has a fitness benefit to both parents, defecting is likely to have a cost. This has been shown by many "widowing" experiments. For example, investigators who removed male Snow Buntings (*Plectrophenax nivalis*) from fourteen nests found that the widowed females could deliver only 73% as much food as control pairs and raised fewer offspring that, overall, weighed 45% less.[29] So defecting males pay a cost: the demise of their genes in dead or less fit offspring.

What determines, conversely, the benefit of defecting and finding other mates, leaving only the other sex to rear the offspring? First, we might expect this option to benefit males more often than females because a female usually can lay only one or a few clutches of large, energy-rich eggs, whereas a male can mate with many females with relatively little effort.[30] This helps to explain why there are more species with female-only than male-only care. Second, the fitness benefit of providing care is lower, and the benefit of defecting is relatively higher, if the nestlings are very likely to be someone else's offspring. (Extra-pair paternity—"extra" meaning "outside"—is thought to occur in about 90% of all bird species.) Likewise, the benefit to a female of providing

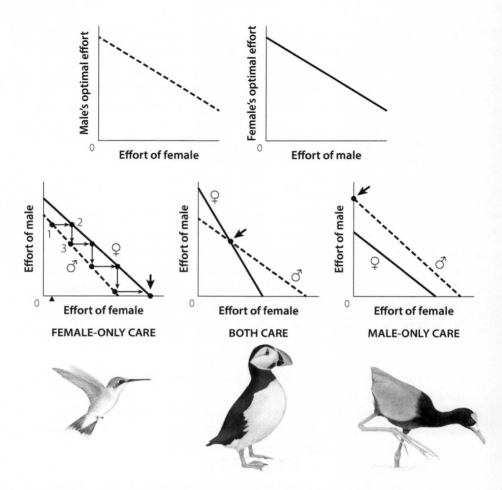

FIGURE 7.4. A slightly mind-stretching model of the evolution of parental care. The more care one parent provides to the offspring, the less care the other need provide, yet still have high offspring survival (and therefore high fitness). The upper two graphs show for each sex that as its optimal level of care (the care needed to successfully fledge all the offspring) declines, the more care the other sex provides. Both lines are plotted together in the lower graphs. At lower left, the female's optimal line is entirely above that of the male. Suppose the population starts at the solid arrowhead on the x-axis, indicating that females make a small effort. For that level of female effort, the optimal male effort is at point 1. Supposing male effort evolves to that point, the new, higher, optimum for females is at point 2, which in turn puts the male optimum lower, at point 3. So we expect the population to evolve, zigzagging down to the large point indicated by the arrow at lower right. This species is expected to evolve all-female care (as in humming-birds). The opposite happens in the lower right graph, resulting in all-male care (as in the Northern Jacana [*Jacana spinosa*]). If the lines cross, the population evolves (in theory) to the crossing point, at which both parents care for offspring (as in Atlantic Puffin [*Fratercula arctica*]). (After Futuyma 2009; art, Luci Betti-Nash.)

parental care is weaker if some of the nestlings were laid by other females: conspecific, or same-species, brood parasitism is known in more than 200 species of birds.

Third, an important benefit of defecting is the chance of having additional offspring by mating again. One factor that could affect this is the sex ratio. In a population with a strongly skewed sex ratio, an individual of the more abundant sex has a lower chance of mating, so it may be advantageous, once mated, to provide care for the young rather than abandon them and face strong competition for additional mates.[31] The rarer sex is expected to provide less parental care because fitness is more easily increased by investing time and energy in searching for more mates.

Another factor is the intensity of sexual selection,[32] which is variation among members of one sex (say, males) in the number of mates they obtain and, therefore, the number of their offspring. The basic idea is that if males vary greatly in mating success—if some mate with multiple females and others with few or none—sexual selection is strong and favors features that increase the chance of repeated successful mating efforts. The time and energy expended on mating efforts come at the expense of caring for offspring. So species in which one sex (often the male) is faced with strong sexual selection are likely to have parental care provided mostly by the other sex—often the female. Often, but not always: in species such as phalaropes, sexual selection is stronger on females, and males provide all the parental care. The same theory applies, but in reverse.

Tamás Székely and colleagues studied the Kentish Plover (*Charadrius alexandrinus*) in southern Turkey, where females, and less often males, were known to frequently desert their brood, leaving care to the other parent. The Székely team found that removing one parent from nests tended by both parents reduced the survival of chicks but that males more successfully defended the chicks, so fewer died than in female-tended nests. When the researchers removed both one parent and the offspring, the now single females found new mates (and laid more eggs) about five times faster (five versus twenty-five days) than the now single males because of the strongly male-biased sex ratio in the population (due to higher mortality of female chicks). Overall, Székely's results support the theory that because of a male-biased sex ratio, females could increase their fitness by leaving offspring in male hands and having multiple broods.

These results suggest how some other plovers, such as the Eurasian Dotterel and the Mountain Plover (*Charadrius montanus*), evolved male-only parental care.[33] Perhaps predominant male care starts to evolve when offspring survival is better under male than female care—as in Kentish Plovers. The ratio of male to female care of offspring is very variable among shorebird species, culminating in the complete sex role reversal seen in species such as phalaropes, jacanas, and painted snipe (plate 25). In these species, females compete for males and are polyandrous (one female may have several mates). Undoubtedly owing to this strong sexual selection, females are larger than males, and in some species are more brightly colored; they do not incubate or care for the young, leaving those duties to the males.

Székely and colleagues compared species of shorebirds that have biparental, female-only, and male-only parental care (figure 7.5). They found evidence for the theory that the more abundant sex would be the caregiver: species with a female-biased sex ratio tend to have "conventional" sex roles with either biparental care or female care. Likewise, males outnumber females in species that have mostly or entirely male parental care.[34] Beyond shorebirds, an analysis of data on 659 species of birds in 113 families supported all the theoretical expectations: species that have a high degree of parental cooperation tend to have low extra-pair paternity, an even sex ratio, and apparently lower intensity of sexual selection.[35]

One way to avoid the fitness cost of offspring care is to arrange for someone else to do it. In the basic model of parental care, that would be your mate, but another option is to lay your eggs in someone else's nest and let them do the work. This is brood parasitism, and it can be conspecific (parasitizing other members of the same species) or interspecific (as in cowbirds, some cuckoos, and a number of other species).[36] Conspecific brood parasitism, usually detected by DNA tests, is known to occur in more than 200 species and seems to be most frequent in waterfowl, grouse and relatives, grebes, rails, swallows, and a few other families. In most such species, brood parasitism in one of several behaviors a female may practice, depending on the situation. Michael Sorenson posited a simple model in which a female's "decision" depends on environmental conditions that affect her physiological vigor.[37] If her condition is bad enough, she doesn't develop eggs; if it is somewhat better, she might develop eggs but have little prospect of rearing the young, perhaps because she cannot find a suitable nest site. An option, then, is to lay them in another female's nest. (This has been called the "best of a bad job" hypothesis.) If a female is healthy and vigorous, she may lay eggs and care for them in her own nest; if she is exceptionally vigorous, she may be able to develop more eggs

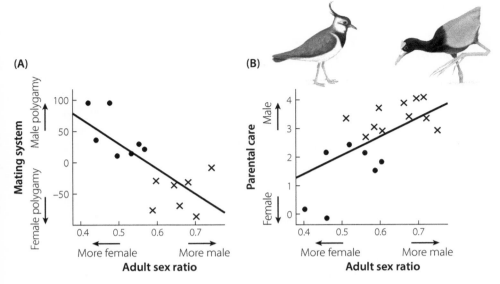

FIGURE 7.5. Mating systems and parental care among species of shorebirds, in relation to female-biased (as in Northern Lapwing [*Vanellus vanellus*]) versus male-biased (as in Northern Jacana) sex ratio. Each point is data for one species. Circles represent species with predominantly female parental care; crosses show species with predominantly male parental care. (A) Species with a female-biased sex ratio tend to have male polygamy (one male has multiple mates), while a male-biased ratio is associated with female polygamy (one female has multiple mates). (B) The mating system difference is mirrored by whether parental care is mostly by females (in species with female-biased sex ratio) or by males (in species with male-biased sex ratio). (After Liker et al. 2013; art, Luci Betti-Nash.)

than she can care for—in which case, she would gain additional fitness by producing offspring that are raised by foster parents.

As an example, the rate of brood parasitism by Eastern Bluebirds (*Sialia sialis*) increased when the researchers, Patricia Gowaty and William Bridges, removed some nest boxes (best of a bad job?). On the other hand, American Coots and Redhead (*Aythya americana*) ducks increased their total fecundity by laying eggs in neighbors' nests, in addition to rearing a nestful of young themselves.[38] In some species, brood-parasitic females seem to assess their neighbors and target the more promising ones. Cliff Swallows nest in very dense colonies, where the major cause of nestling mortality is blood-sucking fleas and swallow bugs (a member of the bedbug family). In the course of their long-term, multifaceted study of this species, Charles Brown and Mary Brown discovered that these birds often parasitize nests near their own. They are somehow able to assess which nests have the lowest load of blood-sucking insects and generally choose the most successful nests.[39] Not surprisingly, Cliff Swallows intensely guard their nests against parasitic females.

Some species, although surprisingly few, can recognize and reject eggs that aren't their own. In the course of his studies of American Coots, Bruce Lyon found that, presumably as a result of strong natural selection, about 40% of females rejected foreign eggs. They used differences in color to distinguish their own eggs from those of interlopers.[40] This is the basis of the evolution of egg mimicry in brood-parasitic cuckoos that I described in chapter 5: cuckoos are successful only if their hosts don't distinguish the parasite's egg from their own. The obligatory brood parasitism of the cuckoos, cowbirds, and others probably originated in the conspecific brood parasitism that so many species display.[41]

Having traversed a range of life histories among birds, and what it takes to rear their young successfully, what about the Malleefowl and other megapodes, with which I started this chapter? You might have been struck, as I was, by how reptilian their nesting is: they bury their eggs and either guard and tend the nest like alligators or abandon it completely, like most other reptiles. Could this be a holdover from the reptilian ancestors of birds?

Phylogeny provides the answer. Look at the phylogeny of birds in figure 3.6 and see that the Megapodidae are one of a number of branches in the Galloanseres (game birds and waterfowl), all of which incubate their eggs with their bodies. A deeper split yet in the phylogeny separates these and most other birds (Neognathae) from the ostrich and other palaeognaths, all of which, likewise, body incubate. So body incubation is ancestral in birds, and the megapodes have evolved external incubation anew. A phylogenetic study of the family suggests that ancestral megapodes built mounds, with males presumably doing the work as in the Malleefowl, and that burrow nesting with no further parental effort, such as in the Moluccan Megapode, evolved later.[42] The first step matches the evolution of male care that we see in some shorebirds. The second, freeing both sexes from ongoing parental care, is equaled only by brood parasites. The megapode life style seems so advantageous, it's surprising that it evolved only once.

8

Auklets' Crests and Peacocks' Trains

SEXUAL SELECTION IN BIRDS

What is it about birds that sparks our interest? I know birders who first were thrilled by the force and majesty of eagles and falcons and some who were enchanted by confiding chickadees or visitors to their bird feeder. But the most common reason, I suspect, is birds' beauty: the brilliant colors of orioles and ducks, the fancy plumes of pheasants and herons, the haunting songs of thrushes and nightjars. As a boy visiting New York's Bronx Zoo, I was captivated by the curled white moustaches and red bill of a slaty Inca Tern (*Larosterna inca*), the iridescence and long white tail streamers of a Ribbon-tailed Astrapia (*Astrapia mayeri*), and the astonishing Wilson's Bird-of-paradise (*Diphyllodes respublica*), with its bare electric-blue crown, brilliant red and yellow back, and circular tail feathers (plate 26).

In many bird species, bright colors, ornaments, and song are male features. North American residents may think of Northern Cardinal (*Cardinalis cardinalis*), Western Tanager (*Piranga ludoviciana*), and most warbler species; Europeans of the Robin (*Erithacus rubecula*) or the Black Grouse (*Lyrurus tetrix*); Australians of a fairywren or a lyrebird. But beauty is in the eye of the beholder, and if the beholder is a female House Sparrow, beauty may reside in a male's black throat and breast; if a female Chipping Sparrow (*Spizella passerina*) or Chiffchaff (*Phylloscopus collybita*), in a male's simple song. In some species, though, the female is the more brilliant sex, as in phalaropes and buttonquails; and both sexes are equally gaudy or ornamented in toucans, parrots, auklets, and many other birds. We would like to understand why birds have evolved these features and why the pattern of sex difference varies.

Darwin—of course—proposed an explanation, developed at length in *The Descent of Man, and Selection in Relation to Sex* (1871). He devoted 241 pages to part 1, "The Descent or Origin of Man," and no fewer than 575 pages to part 2, "Sexual Selection," of which 200 pages are about birds. Darwin distinguished between natural selection (imposed by external environmental factors such as drought or predation) and sexual selection, which "depends on the advantage which certain individuals have over other individuals of the same sex and species, in exclusive relation to reproduction." The key feature is competition among members of a sex for the opportunity to reproduce. Some sex-specific features, such as ovaries and testes, aren't related to competition, and so don't fit this definition. But many other features do. Darwin invoked sexual selection to explain, for example, why males of migratory bird species usually arrive at their breeding areas before females, why males are often larger than females and may have weapons such as the leg spurs of pheasants, and why males (usually) have more exaggerated colors, vocalizations, display behaviors, and ornaments such as plumes and wattles.

Darwin emphasized two forms of sexual selection, as do biologists today. One is direct competition among members of the same sex (usually among males for access to females). The other is the preference of one sex (usually female) for some rather than other members of the opposite sex. This preference results in selection for features that make males more attractive to females. This idea has had a long history of controversy. For decades, biologists doubted that birds or other animals had the ability to choose among potential mates, and Darwin's proposal was neglected until the late 1970s. Now we know that females (of birds, frogs, fish, insects, and probably most other animals) do indeed discriminate among males and do indeed favor males with some characteristics over others. But now there are new controversies centered on the question, why would females prefer males with exaggerated colors, ornaments, or behaviors? What's in it for them?

These two forms of sexual selection might reinforce each other; the Swedish investigator Anders Berglund and colleagues[1] suggested that some features serve both as ornaments (attractive to females) and armaments (aids in male-male conflict). And in a pioneering analysis, Mary Jane West-Eberhard[2] described how sexual selection is often a form of what she called social selection: selection for characteristics that enable successful competition for resources of any kind—food, space, or mates. In all cases, selection is likely to be strong and almost continual, and innovations may provide new avenues to success because competitors aren't adapted to deal with them. The result, she suggested, is that characteristics used in competition may evolve to be more and more extreme and may also come to differ between related populations or species in which mutation may provide different innovations. This process might even cause speciation—evolution of a species into two or more different species. (More on that in chapter 10.) Moreover, social selection might explain why females of some species have elaborate ornaments.

We now have evidence for just about everything Darwin suggested about male-male competition. Male Pied Flycatchers (*Ficedula hypoleuca*) that arrive at the breeding area earlier in the spring have more offspring, just as Darwin supposed. They may acquire better territories or perhaps have better genes.[3] Males of many species compete aggressively for females directly or for territory that will include females and the resources needed to rear offspring. Competition can include physical combat, aided in a few species by weapons such as the wrist spurs of the Spur-winged Goose (*Plectropterus gambensis*) and of screamers (figure 8.1)[4] and the leg spurs of some gallinaceous birds, such as chickens and other members of the pheasant family.[5] In many bird lineages, the advantage of larger size in male conflict has resulted in the evolution of larger males than females. Aggressive encounters between males are almost unremitting in species that mate in leks: sites where males gather and defend small territories against each other. Females visit the lek and choose one or more males, often dominant males that hold territories near the center. Lekking species include several species of grouse (such as prairie chickens), sandpipers (such as the Ruff), and some tropical rainforest birds (e.g., manakins, cock-of-the-rock). Their aggressive displays often feature ornaments, such as the extravagant plumes of some birds-of-paradise, the head plumes and vocal sacs of prairie chickens, and the Ruff's eponymous ruff of long feathers (see figure 5.1). In many lekking species, only a few of the males successfully mate with visiting females, so sexual selection is intense.[6] In these species and others that do not form pair bonds, males provide no help to females—only their sperm.

The features displayed during aggressive encounters often serve as "badges" of status that may confer mating success. Male Yellow Warblers (*Setophaga petechia*) vary in the amount of reddish-brown streaking on the breast. Mounted birds, placed in active territories, elicited more aggressive reactions from territorial males if they had more streaks, suggesting that they were perceived as a greater threat.[7] Likewise, male House Sparrows with larger black badges obtained mates earlier in the spring, and females responded more strongly (by soliciting copulation) to mounted male specimens with larger

FIGURE 8.1. A Horned Screamer (*Anhima cornuta*), showing the wrist spurs. The horny spur coverings have been found embedded in the breasts of other screamers, showing that they are serious weapons. (Art, Stephen Nash.)

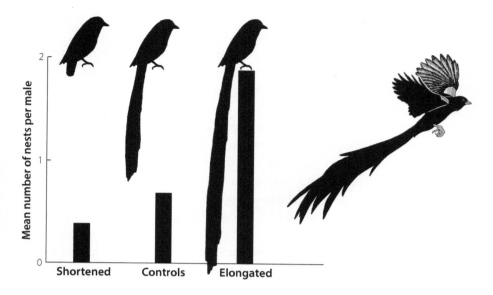

FIGURE 8.2. The mating success of control and altered male Long-tailed Widowbirds (*Euplectes progne*). The number of females that nested in a male's territory was, on average, lower than for control (unaltered) males if their tail feathers were shortened but dramatically higher if the feathers were lengthened well above the normal length. The lengthened tail may be a "supernormal stimulus," a concept mentioned in chapter 7. (After Andersson 1982 and Futuyma and Kirkpatrick 2017; art, Stephen Nash.)

badges. Their preference might have been advantageous because males with larger badges held territories with more and apparently better nest sites.[8] This is an example of Anders Berglund's suggestion that a male feature may be advantageous both in competition with other males and in its appeal to females.

Darwin's hypothesis that females exert sexual selection for exaggerated male traits has been thoroughly verified. One of the most celebrated examples is Malte Andersson's experiment with Long-tailed Widowbirds (*Euplectes progne*) in Kenya (figure 8.2).[9] The male has tail feathers three to four times its body length that droop vertically during his display flight above his grassland territory. This species is polygynous (Greek for "many females"): successful males commonly have several mates. Andersson shortened the tails of some males and attached the clipped ends to other males, lengthening their tails well beyond the normal range. Compared with normal males (with cut and reattached tail feathers, so controlling for any effects of the handling process), those with shortened tails attracted fewer females, and those with abnormally long tails attracted about twice as many.

The poster boy of sexual selection is surely the peacock (more formally the Indian Peafowl, *Pavo cristatus*), renowned for his train of long back (not tail) feathers, ornamented with eyespots, that are erected and spread during courtship. Studying a free-ranging population in a large English park, Marion Petrie and her coworkers found that females mate more often with males with longer trains and more eyespots. (And when she clipped off some eyespots, males attracted fewer females.) Some males are therefore more popular than others, and Petrie found that a female who mates with a less popular male will often mate again, with a more popular one. Other investigators found that males' mating success was also related to their level of display activity and the color and iridescence of their eyespots.[10]

In species that are polygynous (like the widowbird) or promiscuous (like the peafowl), some males have multiple mates and others few or none—and so there is sexual selection for whatever features make males more successful. But what about species that form pair bonds? If the population sex ratio is 1:1, and every male gets a female, it would seem that there wouldn't be any advantage to having a longer tail or brighter colors. How do we account for the bright male coloration of the Blackburnian Warbler, the Bluethroat (*Luscinia svecica*), and countless other pair-bonding species?

At least part of the answer is that such species may be "socially" monogamous but not sexually monogamous: in many species, both sexes engage in EPC: extra-pair copulation (adultery).[11] This is now easily documented by DNA paternity tests, just as in humans. (In some species, the level of extra-pair paternity is shocking, as in the Australian fairywrens, jaunty, long-tailed birds that live in cooperative groups that may include a female and several males that all help to rear her offspring. A study of the Superb Fairywren [plate 2] revealed that 76% of offspring were sired by males outside the group.[12] Female fairywrens even travel across intervening territories to mate with especially popular males.) Extra-pair copulation contributes to sexual selection in socially monogamous birds. A study of Collared Flycatchers in Hungary found that males with a wider white patch on the forehead sired more offspring by EPC, while "cuckolded" males with narrower patches fathered fewer offspring. Females whose social mates had narrow patches were much more likely to engage in EPC with more attractive males. This behavior selects for wider forehead patches. A similar result was found in Tree Swallows in the United

States: brighter blue males have more extra-pair offspring than males that are naturally dull or were made duller by inking their feathers.[13]

(A digression: Why has EPC evolved? Its advantage to males is clear: an allele that inclines its male carriers to mate with many females automatically increases in frequency. Why females engage in EPC is less obvious. An intriguing possibility is that female extra-pair mating behavior is a by-product of the genes that incline males to do it. After all, these genes are inherited by both sons and daughters. Some evidence for this idea is that the sisters of male Zebra Finches that were more prone to court unfamiliar females were also more strongly inclined to mate promiscuously. There are also several adaptive hypotheses. Females might seek more vigorous males than their own social mates if those males, even if already paired, are likely to sire more vigorous offspring. This proved to be exactly the case in the Blue Tit, in which "extra-pair young" were found to have higher survival than "intra-pair young."[14])

Darwin inferred that the advantage of bright colors and other male ornaments must outweigh some disadvantage because in many species the females are cryptically colored and harder to see and males moult into a drab plumage after breeding. The peacock's great mass of long train feathers impairs its ability to take flight, and the long tail feathers of certain hummingbirds increase the metabolic cost of flight.[15] Even aside from the disadvantages of such male features, the big question, the one that dominates research on sexual selection, is why females prefer these features in males—and therefore cause the features to evolve. Why have females evolved these quirky preferences? I'll describe the three most-discussed proposals.

The first is that females don't gain anything by their quirky preferences. This is the position taken by, among others, Michael Ryan at the University of Texas, whose perspective combines evolutionary biology and neurobiology. In *A Taste for the Beautiful*,[16] Ryan writes (p. 3), "instead of the brain having evolved to detect beauty, the brain determines what is beautiful . . . beauty is in the brain of the beholder." This might be for several reasons. Some signal characteristics are more easily perceived in relation to the environment. For example, songs of open-country birds tend to be higher pitched than those of forest birds because different frequencies travel better in these contrasting conditions. The color and brightness of face and breast marks that serve as social signals differ between canopy-dwelling and understory-dwelling species of rainforest birds, having evolved to be readily detected in the species' typical environment.[17]

Stimuli, once sensed, are interpreted by the brain, and a species' brain may have idiosyncracies that cause a bird to respond to novel characteristics. Both sexes of Crested and Whiskered Auklets (*Aethia cristatellus, A. pygmaea*) have facial ornaments, which are favored by the opposite sex. Other species of auklets lack crests and whiskers, which are a derived ("advanced") characteristic within the auklet clade (figure 8.3). Ian Jones and Fiona Hunter attached crest feathers of Crested Auklets to the forehead of Least Auklet (*A. pusilla*) specimens that were mounted in a realistic position and placed these models in a Least Auklet colony. The resident birds reacted with sexual displays about ten times more frequently to the models with artificial crests than to "normal" models that lacked them! So the Least Auklet has an ingrained preference for a crest even though—according to the phylogeny—it has never had a crested ancestor.[18] If this was also true of the crestless ancestor of the Crested Auklet, it is likely that when mutant crested males first occurred, they were instant objects of desire, and their higher reproductive fitness was assured. Along the same line, Nancy Burley attached artificial crests—single large feathers—to the head of captive Zebra Finches and Long-tailed Finches (*Poephila acuticauda*) and discovered that both species preferred mates with crests, compared with normal crestless birds. None of the 120 species in this bird family (Estrildidae) have crests, but they seem to have "veiled" preferences that can provide a reproductive advantage to a novel trait.[19]

Sensory biases of this kind certainly can't account for the great variety of preferences that diverse species have for different male traits, nor for all their details, but they might account for the first steps in their evolution. There are two other major theories that might account for the great variety of sexually selected characteristics.

Probably the most popular idea, among researchers in this field, is that females get some kind of benefit, directly or indirectly, from choosing males on the basis of one or more display characteristics. A more pronounced male feature—color, tail length, song, display behavior—might signal that the male would directly benefit the female or their offspring, perhaps by providing more food or better defense against predators. Or such males might transmit better genes to a female's offspring, enhancing their vigor or ability to resist parasites and disease. The choosy female would benefit indirectly, by having offspring with "good genes." Whether the benefit be direct or indirect, the male's feature is an indicator—it is correlated with his vigor or "genetic quality." This correlation might exist if genetically "healthier" individuals are more active or can devote more energy to developing the display feature. The Israeli

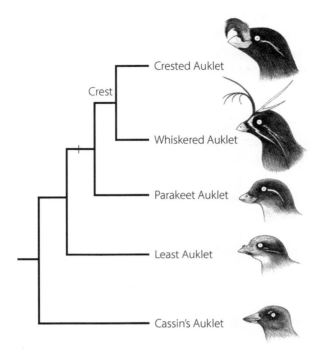

FIGURE 8.3. The phylogeny of the five species of auklets. The crest
evolved in the common ancestor of the upper two species. As far as
can be determined, the Least Auklet never had an ancestor with a
crest, so there never was any selection for females to be attracted to
a crested mate. Nevertheless, Least Auklets reacted enthusiastically
to specimens with attached crests. (Art, Stephen Nash.)

biologist Amotz Zahavi suggested that a costly male ornament is a handicap
and that a male's highly developed ornament is evidence that he is genetically
healthy and vigorous enough to have developed it, despite its cost.[20] Every
individual in a sexual species, whether bird or human, carries some rare harmful mutations. (They are the subject of most medical genetic and genomic
research.) Individuals with fewer such mutations may be more vigorous, better
able to develop pronounced display features, and more likely to father healthy
offspring.[21]

There is considerable evidence for the indicator, or "good genes," idea. The
intensity of the red plumage of male House Finches is correlated between
fathers and sons, so it has a partly genetic basis. Geoffrey Hill showed that

females prefer to mate with brighter males and that these males feed their mates and nestlings more often. He concluded that plumage coloration is a reliable indicator of male parental care. In a later study, Hill found that redder males are more resistant to the bacterium that causes conjunctivitis and death in this species: they cleared experimental infections faster.[22] This study supported a hypothesis proposed twenty-three years earlier, when William Hamilton and Marlene Zuk suggested that sexually selected male characteristics are indicators of a better ability to resist parasites and disease.[23]

Research groups led by Marlene Zuk and David Ligon have studied captive populations of Red Junglefowl (*Gallus gallus*), the ancestor of domestic chickens (plate 27). Surprisingly, the male's elaborate plumes and bright colors seem not to influence females' choice of mate.[24] Instead, the key sexually selected feature is the comb: males with larger, brighter red combs dominate other males and are preferred by females. They have better body condition, as well as higher testosterone level and stronger immune response. Comb size is genetically variable. Timothy Parker artificially inseminated hens with sperm from various males and found that the sons of large-combed males inherited large combs and had higher body condition. The evidence is consistent with the idea that a large comb is an indicator of "good genes" that would make a male a desirable mate.[25]

Another example is the Great Reed Warbler. Investigators in Sweden who studied a population for six years found, by DNA testing, that 6% of broods included at least one nestling that resulted from extra-pair mating.[26] The genetic fathers of these offspring were those neighboring males with the largest repertoire of different songs—larger than the female's social mate's repertoire. In accord with the good genes model, the post-fledging survival of offspring (measured by whether or not they returned the following spring) was correlated with their genetic father's song repertoire.

A third hypothesis for the evolution of female preferences has a noble pedigree but is very controversial. It is usually called "runaway sexual selection," or sometimes the "sexy son hypothesis." It was proposed in 1930 by Sir Ronald Fisher, a mathematically talented British biologist who pioneered and largely shaped modern evolutionary biology by synthesizing Darwin's concept of natural selection with the science of genetics. In this model, the evolution of a female preference and a male trait, once it starts, becomes a self-reinforcing process.[27]

Fisher's model starts with rare alleles for a male ornament and also rare alleles (of a different gene) that incline females to prefer males with the ornament. If rare ornamented males attract more females than the average nonornamented male, their sons will inherit the appealing trait and will likewise be more successful in mating, so ornament alleles will increase slightly in the population. But the mothers of many of those sons carried and passed on their preference alleles, so the greater reproductive success of ornamented sons increases the frequency of both the ornament alleles and the preference alleles.[28] As the frequency of preference alleles increases, more females prefer ornamented males, which then have an even greater reproductive advantage than before—and the process repeats, causing the frequency of both ornament alleles and preference alleles to increase further and faster—a snowballing process.

This process may result in ever more extreme female preference and more elaborate male features—until they have such a countervailing disadvantage that the process stops. The disadvantage is that sexually selected male characteristics—bright colors, elaborate plumes, loud songs, the energetic demands of courtship and male rivalry—are likely to reduce survival. (An extreme case, although not in birds, involves the accidental human introduction of a parasitic fly to the Hawaiian Islands, where the fly homed in on the male call of a species of cricket. Within twenty generations, the crickets evolved modified wings that do not produce song.[29])

I think Fisher's sexy son hypothesis is a really neat concept, but it isn't winning the contest of ideas. For one thing, it rests on mathematical theory, and the match between theory and real life depends on what assumptions were built into the theory. (For an example that is topical as I write: an epidemiologist may model how the spread of a virus may be stopped, but the model won't work if it assumes that everyone is willing to be vaccinated and a significant percentage of people actually refuse.) The early mathematical models made the simplifying assumption that genetic variation in female preference has no intrinsic cost or benefit and just rides along with the male trait by genetic hitchhiking (cf. chapter 5). More advanced models assume that choice is costly; females may lose reproductive opportunities if they bypass males in what may be a frustrated quest for the ideal mate.[30] In these models, the runaway process is hampered and may not occur at all.[31]

Several researchers who studied insects, frogs, and other animals reported that sexy fathers have sexy sons, which is consistent with Fisher's idea[32] but isn't enough to show that female preference evolves for that reason. The ornithologist Richard Prum enthusiastically advocates the runaway hypothesis in his book *The Evolution of Beauty*,[33] but stronger experimental evidence will be needed to resolve the issue. At this time, most researchers in this area are inclined to think that sexual selection of males by females may best be explained by sensory bias or by benefits to the female or her offspring.

Now let's step back and ask a fundamental question: why have we been so focused on male characteristics? Don't females compete for mates? Wouldn't some females make better mates than others and be preferred by males? Doesn't sexual selection apply to females as well as males?

Yes, definitely. That so many species have more brightly, highly ornamented males compared with females is a result of stronger sexual selection on males, but this isn't universal. As I described in chapter 7, parental care is the male's duty in some species, such as phalaropes. Since he can care for a limited number of offspring, from one female or a very few, there can be more variation in reproductive success among females, which often mate with and lay eggs for several males. Probably because they experience stronger sexual selection, females court males and are more brightly colored.

But females also have pronounced coloration, plumes and other physical ornaments, or song in many species that don't have reversed sex roles. In some, these features are subdued versions of the male's trait; in others, the sexes are equally adorned. Three possible explanations of female ornaments have been suggested. One was proposed by Darwin himself, who thought it "probable that the ornaments common to both sexes were acquired by one sex, generally the male, and then transmitted to the offspring of both sexes."[34] Today, with our understanding of genetics, we would say that the female's feature is a side effect, or correlated effect, of the genes that underlie the male's feature.[35] (A reverse example, in humans, might be men's nipples.) For instance, there is a perfect genetic correlation between male and female Pied Flycatchers in the size of the sexually selected white forehead patch: related birds of both sexes have similar patches.[36] Possibly the subdued male-like features of some female birds are nonadaptive genetic side effects. Among many possible examples is the Tufted Duck (*Aythya fuligula*), named for the long drooping feathers on the back of the male's head. Females bear a very small tuft that I imagine could be a genetically correlated side effect (figure 8.4).

But most female ornaments are thought to be functional and to have evolved either by sexual selection (as in males) or by other forms of social selection (as Mary Jane West-Eberhard proposed in 1983).[37] We now know that in many species, males choose among females just as females choose among males, and in some cases, males obtain direct or indirect benefits. Male Barn Owls prefer females that have larger, darker black spots. These females lay

FIGURE 8.4. Male and female Tufted Ducks (*Aythya fuligula*). The female has a slightly developed version of the male's conspicuous crest. There is apparently no evidence of whether or not the female's crest has any function. (Art, Luci Betti-Nash.)

larger clutches of eggs, their offspring are more resistant to blood-feeding flies, and more of their offspring survive to fledging. So the males' fitness is affected by their choice of females.[38]

In some species, female features have evolved mostly because of social selection. The sexes look much alike in most parrots, but the Eclectus Parrot

(*Eclectus roratus*) of northern Australia and New Guinea has extraordinary sexual color dimorphism: the male is green and the female is brilliant red and blue (plate 28). Females compete strongly (and viciously) for nesting holes in tree trunks and spend most of their time using and defending their holes. Their conspicuous color is thought to advertise ownership of their precious, scarce resource. Males compete for females, and several males attend a female, providing her and her young with food. Males' coloration matches the foliage in which they spend most of their time foraging. In the Greater Vasa Parrot (*Coracopsis vasa*) of Madagascar, both sexes are black, but females develop bright orange skin on their head while they rear chicks, which may have had up to three different fathers. The males feed the female, who then feeds the chicks. Females intrude into each other's trees, uttering loud, complex songs as they compete for food-carrying males. The researchers suggest that their head color and complex song evolved by social selection.[39]

In many species, both sexes have similar ornaments, song, and displays. Cranes, herons, and many seabirds, such as albatrosses, penguins, and gannets, are well known for their elaborate mutual displays and greeting ceremonies that seem to enforce their pair bond. In some cases, there is clearly mutual sexual selection. Ian Jones and Fiona Hunter studied Crested Auklets, in which both sexes have a long, forward-curving crest on the forehead (figure 8.3).[40] They presented birds with models (stuffed specimens in realistic positions) that had crests of different lengths. Males were less aggressive toward models with longer crests, which evidently signal dominance status. Both males and females made more frequent, longer sexual displays toward opposite-sex models with longer crests. The experiment shows that both forms of sexual selection (same-sex aggression and opposite-sex preference) favor longer crests in both sexes. In birds like this, in which the survival and health of offspring depends on good care from both parents, both sexes would do well to choose a high-quality mate, perhaps indicated by its ornaments.

Many passerine species that hold year-round territories in tropical forests, such as many wrens, antbirds, babblers, and flycatchers, show the same pattern. These species often have low rates of mortality and hold a territory for several or many years, and they are generally socially monogamous and engage in little extra-pair copulation. Joseph Tobias and colleagues studied two species of warbling antbirds[41] in a Peruvian rainforest, noting when they sang throughout the year, how they responded aggressively to playback of their songs, and how they reacted when an individual's mate was temporarily removed. These species live in dense undergrowth and use vocal signals more than visual signals such as coloration. The researchers found that pairs re-mained together throughout the several years of their study, and both sexes sang throughout the year, not just during the breeding season. Both sexes de-fended their territory, responding aggressively to playback, especially of same-sex song. They also sang more if their mate was removed, suggesting that they use song to attract mates. Sexual selection and social selection both seem to have been important.

Does evolution always lead toward brighter colors and more exaggerated ornaments? In at least some cases, the trend has been reversed. The aptly named Sombre Hummingbird (*Eupetomena cirrochloris*) is one of several dull-colored twigs on branches of the hummingbird phylogenetic tree that are otherwise populated with brightly ornamented species. The fifty or so species of the tanager genus *Tangara* are among the most diversely and strik-ingly colored Neotropical birds, bearing names such as Paradise, Opal-rumped, Golden, Emerald, Seven-colored, Brassy-breasted, Beryl-spangled, and Flame-faced Tanager (plate 29)—but nested among them is the drab Plain-colored Tanager (*Tangara* **inornata**). The phylogeny implies in both cases that the dull species have evolved from more gaudy ancestors.[42] There is some evidence, based on museum specimens, that male coloration in Eu-ropean Black-tailed Godwits (*Limosa limosa*) has become progressively duller since 1840, and that duller males have higher fitness.

Not much is known about why birds might evolve to be less ornamented, but an interesting pattern has been found in phylogenetic studies of New World orioles and warblers.[43] In both groups, tropical species are sexually monomorphic—both sexes have the same bright color patterns—but fe-males are less highly colored than males in migratory temperate-zone species. The phylogenies, based on DNA sequences, show that both families were ancestrally nonmigratory and sexually monomorphic and that migration has evolved several times in each family, always accompanied by the parallel evo-lution of a drab female plumage (figure 8.5). We have seen that in many non-migratory tropical birds both sexes defend territories year-round and have socially selected characteristics. The resulting social selection of females is probably alleviated when a species becomes migratory, and other factors, such as the risk of predation, become relatively more important.

I've described several forms of sexual and social selection that help to an-swer the question of how we can explain exaggerated ornaments like the pea-cock's train. There is one more intriguing twist on this theme: the idea that sexual selection can arise from conflict between the sexes. This was proposed

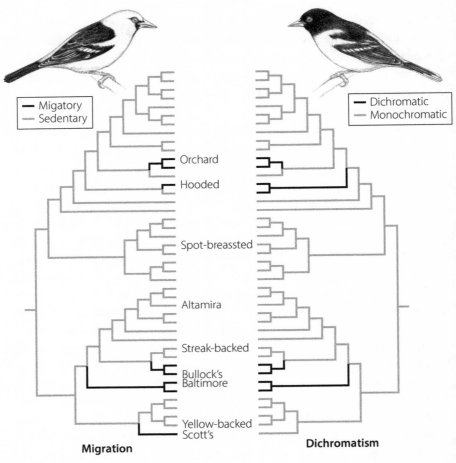

Migratory
Sedentary

Dichromatic
Monochromatic

Orchard

Hooded

Spot-breassted

Altamira

Streak-backed

Bullock's
Baltimore

Yellow-backed
Scott's

Migration

Dichromatism

FIGURE 8.5. A molecular phylogeny of the American orioles (genus *Icterus*), portrayed twice as mirror images. The black branches show migratory lineages in the left figure and sexual color dimorphism in the right figure. Most species are tropical and have the ancestral state: nonmigratory and with the sexes having the same coloration. The correlation isn't perfect, but most of the lineages that evolved a migratory habit also evolved a sex difference in color pattern. For clarity, the English names of only a few species have been entered. Above the diagram, a nonmigratory Altamira Oriole (*Icterus gularis*, left) and a migratory male Baltimore Oriole (*I. galbula*, right). (After Friedman et al. 2009; art, Stephen Nash.)

by Brett Holland and William Rice, based on evidence that in fruit flies—get this—something in the males' seminal fluid enhances the success of their sperm but is harmful to females! Holland and Rice actually found that males in laboratory populations of fruit flies evolve to become more harmful and that females react by evolving higher resistance. Many species, they suggested, might have some form of such sexual conflict, and this might cause "chase-away" selection, in which females evolve to resist males and males evolve to overcome female resistance. We could imagine that males evolve to be so attractive that females mate too readily—at the wrong time or with males that will be poor parents despite their seductive plumage. Females would be under selection to be more resistant and discriminating—and this would renew selection on males to be even more alluring.[44]

One of the clearest examples of sexual conflict in birds is the reproductive anatomy of ducks, described by Richard Prum in a fascinating chapter of his book, *The Evolution of Beauty*. The reptilian ancestors of birds surely had a penis (or phallus), as do all living reptiles. It was lost during the ancestry of most birds (Neoaves) but was retained by the earliest branches of the phylogenetic tree of living birds, including palaeognaths (ostriches, cassowaries, tinamous, etc.) and waterfowl (ducks, geese). Prum describes the astounding penis of the Andean Duck (*Oxyura ferruginea*, closely related to the Ruddy Duck, *O. jamaicensis*), in which a 30-centimeter male has a 41-centimeter phallus. Moreover, it is spirally twisted and has ridges and bumps (as in many reptiles) that are probably an adaptation to keep it lodged in the female's vagina. Ducks' vaginas are also spirally twisted (but in the reverse direction!). Patricia Brennan and colleagues, in a comparison of various waterfowl, showed that in species with a longer, twistier penis, females have a more complex vagina, with more twists and dead-end pockets.[45]

Why?! The answer seems to be "forced extra-pair copulation." In Mallards (*Anas platyrhynchos*) and many other ducks, males often aggressively pursue, subdue, and mate with females that try desperately to escape. Sometimes several males will pile onto a female, and there are even records of females drowning as a result. Christopher Coker and colleagues studied the male genitalia of fifty-four waterfowl species in relation to information on the breeding system.[46] Some species are monogamous, some are promiscuous, and some have a "mixed mating strategy": males maintain a pair bond with one female but also engage in forced extra-pair copulation. The Coker team found that species with higher rates of forced extra-pair copulation have larger testes, a longer penis, and more bumps and ridges. They proposed that these species have evolved a longer penis to prevent the females from shaking loose or to prevent other males from dislodging them. Patricia Brennan and Richard Prum suggested the vagina's dead-end pockets and clockwise spiral coil "function to exclude the intromission of the counter-clockwise spiraling male phallus without female cooperation."

I won't venture to guess if you find the evolution of duck genitals as intriguing as I do, and I admit that I didn't expect a chapter that opens with the beauty of birds-of-paradise to end with sexual conflict. But conflict is a major theme in evolutionary biology, inherent in the very concept of natural selection. I will sound a possibly more pleasing tone in the next chapter, on cooperation.

9

Anis, Swallows, and Bee-eaters

THE SOCIAL LIFE OF BIRDS

Many species of birds can be seen in groups. Wherever you live, you can see some species that are characteristically in flocks, whether small (many jays, tits, scrubwrens) or large (migrating flocks of swallows and shorebirds, winter flocks of hundreds or thousands of Red-winged Blackbirds [*Agelaius phoeniceus*] or European Starlings). You may have seen mixed-species flocks of migrant warblers or winter groups of chickadees or tits with woodpeckers and nuthatches. You may have walked for an hour in a Neotropical forest that seems devoid of birds and then happened on a foraging flock of antwrens, antshrikes, flycatchers, woodcreepers, spinetails, and diverse other species. And some species—Cliff Swallows in North America, Common House Martins (*Delichon urbicum*) in Europe, and many waterbirds—nest in colonies. Why do birds socialize? And how do their social behaviors evolve?

A bird that joins a group always incurs a potential cost. Birds in groups often compete for food; they may have a heightened risk of contracting a disease or parasite; and there is often direct conflict. For example, Acorn Woodpeckers (*Melanerpes formicivorus*) live in social groups, with several females that all lay eggs in the same nest hole. A female's first-laid eggs are usually destroyed by other females, a process that ends only when all the females lay on the same day.[1] So we must ask what potential benefits might offset the cost. This question is even more pressing if a bird's social behavior involves cooperating, or helping other birds, at some fitness-reducing cost in time or energy.

If you are tempted to think that social behavior and cooperation might evolve because it is good for the population or the species as a whole, please resist the temptation. Recall that evolution by natural selection proceeds if an allele that affects a characteristic (such as a behavior) enhances its carriers' survival or reproduction, relative to a prevalent alternative allele. Suppose all members of a population have an altruistic, self-sacrificing genotype. A "self-ish" mutation, one that disposes its carrier to be a cheater or freeloader (what social scientists call a free rider), would reap benefit without paying cost and would increase and possibly replace the "cooperation" allele. Natural selection among genes or individual organisms isn't affected by the benefit or cost to the entire population. By analogy, "what's good for General Motors" isn't necessarily "good for the country."[2]

Nevertheless, birds are, overall, at least as social and cooperative as any other class of vertebrates.[3] Even the cooperation between mates that together care for their young is an advanced form of social behavior that is rare in other animals. Researchers who have devoted their careers to studying social birds suggest that four features of birds incline them to form social groups.[4] Species that fly have wide latitude in choosing where to feed and breed, sometimes in different places, which can lower the cost of cooperating. Birds maintain a high body temperature, so it costs a lot of energy to rear offspring. Both females and males often can contribute to rearing offspring (unlike mammals, in which only females provide milk). And males tend to disperse less from their birth place than females, so compared with other animals, they are more likely to live near their relatives.

Not all social behavior involves cooperation and self-sacrifice; birds often form groups for purely selfish reasons. The behavior of group members often provides "public information" about food. I have often seen Northern Gannets (*Morus bassanus*) stream toward a feeding site from a great distance, having spotted other gannets actively diving and feeding. In the course of his comprehensive research on Cliff Swallows in Nebraska, Charles Brown found that individual swallows obtain more food by using two kinds of information from other colony members. First, individuals catch more insects if they are in a foraging flock than if they forage alone, and solitary individuals join flocks if they have a lower capture rate than flock members. Second, Cliff Swallows nest in dense, often very large colonies, and the colony is an "information center." Brown discovered that swallows that have foraged unsuccessfully return to the colony, locate a successful incoming individual, and follow that bird back out to the productive food source.[5]

A very important advantage of being in a group is "safety in numbers," sometimes called the "selfish herd" principle. Being in a flock, whether of a single species or multiple species, provides protection from predators. The greater the number of potential targets, the lower the chance an individual has of being caught ("dilution effect"); and the melee of alarmed birds in flight can confuse the predator and reduce its chance of success altogether. The best place to be, if a predator is spotted, is in the center of the flock ("better it get you than me!"). That's why starlings, sandpipers, and some other birds take to the air in compact, swerving flocks when they see a falcon (plate 30). Will Cresswell, at Edinburgh University, found that Common Redshanks in Scotland were less likely to be killed by Peregrine Falcons (*Falco peregrinus*) and Eurasian Sparrowhawks (*Accipiter nisus*) when they were in larger flocks, especially if they were in the densest part of a flock.[6]

Many flocking species emit alarm calls when they detect danger, and this reduces their companions' risk.[7] Michael Griesser studied groups of Siberian Jays (*Perisoreus infaustus*) that included both breeding and nonbreeding members and exposed them to models of Northern Goshawks (*Accipiter gentilis*). In most cases, breeders called in alarm, and some breeders were consistently more likely to do this than others. Nonbreeders that were warned reacted to the hawk model faster and had higher survival through the winter than those that weren't warned. And in many species, flock members spend less time being vigilant, and more time feeding, than when they are alone, since the flock members collectively are more vigilant and will call if they perceive danger.

Birds enjoy these and other advantages in both single-species flocks and mixed-species flocks.[8] For example, some species may know the location of food sources or flush insects that others capture. In tropical American forests, swarms of army ants prey on insects, spiders, and even small vertebrates, many of which are flushed as they attempt to escape but then are food for diverse species of birds, some of which are "obligate" ant followers. The calls of birds at a swarm attract other birds, which respond especially to the obligate species' calls.[9] In Africa, the Fork-tailed Drongo (*Dicrurus adsimilis*) scares off certain other species by mimicking their alarm calls and then steals their food.[10]

Alarm calls usually benefit their recipients—but (apart from the drongo) what do the callers get out of it? They are at risk of drawing the predator's attention to themselves, so there is probably a compensating fitness benefit.

Could alarm calls directly benefit the caller? One possibility is that the bird is telling the predator, "I see you, you can't take me by surprise, so don't waste your time; go away." Under this hypothesis, the call isn't directed at the bird's companions, but they nevertheless recognize that it means danger. There doesn't seem to be much evidence for this idea in birds. But Will Cresswell, studying Common Redshanks, found another advantage to the caller: it alarms other birds, which join it to form a tight, flying flock, a "selfish herd."[11]

This brings to mind "mobbing" behavior. You may have seen small birds harassing crows or hawks, diving at them in midair. It isn't quite as dangerous as it looks because a small, close bird can maneuver away quickly. Birds also mob sitting predators such as owls, giving alarm calls that attract other birds that may dilute the danger. The predator has no hope of catching a bird by surprise, and it often will leave, briefly pursued by the noisy mob. (Wherever pygmy owls occur, you can quickly lure in diverse small birds by imitating one or playing its call. And in some parts of the world, you can attract woodland birds by "pishing," which may sound like a mobbing call.)

Still, it's not clear that individuals always, or even usually, benefit from giving alarm calls. So let's consider the possibility that alarm calling is at least slightly altruistic and ask how, in general, altruism could evolve. Three major ways seem likely.

The first is "reciprocity": I'll give you something if you give me something in return. This can work only if the parties repeatedly interact, recognize each other, and remember at least their recent encounters. The best example that I know of in birds is owl-mobbing behavior in the European Pied Flycatcher (figure 9.1). Indrikis Krams and colleagues chose trios of occupied nest boxes, randomly labeled A, B, and C. They placed a stuffed owl near box A and allowed nesting pairs of flycatchers from boxes A and C to mob it—but they prevented box B birds from coming out to help. Some time later, they placed the owl near box B or C. They found that birds from box A readily mobbed the owl if it was near the C box but not the B box: they aided only the birds that had helped them before. This research team also found that breeding pairs generally help near neighbors mob an owl mount but that they usually assisted a distant pair only if that pair had previously helped them.[12]

The second possible reason to cooperate is delayed benefit. Male manakins in the genus *Chiroxiphia*, such as Long-tailed and Lance-tailed Manakins (*C. linearis, C. lanceolata*), are strikingly colored in black and blue with a red crown. They perform a cooperative courtship display at a permanent site in tropical American rain forest (figure 9.2). A dominant "alpha" male and one or more subordinate ("beta") males, in a multiyear partnership, court females by performing an elaborate dance in which each leapfrogs over the others as they sidle along a branch. With very rare exceptions, betas are sidelined, year after year, and only alphas get to mate. Why do betas cooperate?

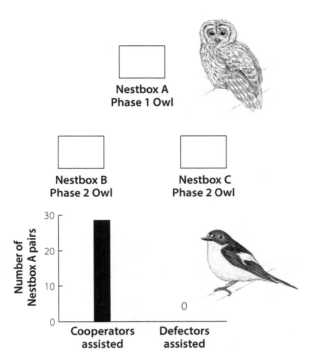

FIGURE 9.1. A test of reciprocity in the Pied Flycatcher (*Ficedula hypoleuca*). In phase 1, a stuffed owl was placed near nest box A. Birds in nest box C, labeled "cooperators," were allowed to help A-box birds to mob the owl; birds in nest box B, labeled "defectors," were prevented from helping. This test was repeated with many trios of boxes. Later, in phase 2, the owl was placed near either a C box or a B box. The columns show that many A-box pairs assisted former cooperators (C-box birds) to mob the owl but did not assist former defectors (B-box birds). Very few A-box birds remained at their nest without assisting. The A-box birds remembered who helped them earlier and returned the favor. (After Krams et al. 2008; art, Stephen Nash.)

Probably because they have no choice: females respond only to multiple, not single, males, so a young male has to join an older, established alpha. Eventually, the alpha dies, and a beta can assume his position and status. In a five-year study of Lance-tailed Manakins, Emily DuVal found that a beta often replaces his long-time dominant partner and inherits the display site, but some betas become an alpha at another site, joining the resident beta.[13] DuVal postulates that their long association with experienced alphas helps betas to learn

FIGURE 9.2. The courtship display of teams of male Lance-tailed Manakins (*Chiroxiphia lanceolata*). The males more or less rotate in sidling along a branch, then jumping over the others and repeating the process. (Art, Luci Betti-Nash.)

and become proficient in displaying. Beta males aren't thinking ahead (as far as we know) or plotting strategy; they are performing behaviors specified by genes that gave their ancestors a reproductive payoff.

The most common basis for evolving altruistic behavior is helping relatives. Recall the simple reason that parental care of offspring is advantageous: a mutation that disposes a parent to care for its offspring will be inherited by at least some of those offspring, so the mutation is, in effect, helping other copies of itself to survive. By the same token, a mutation that disposes an individual to help siblings survive and reproduce can sometimes spread through a population because they have an appreciable chance of having inherited that same mutation. Each individual has two copies of a gene, one from each of its parents. If I have a "helping" allele from my mother (or father), the chance that my daughter or son will inherit (and share) that allele is one-half. The chance that my sister (or brother) inherited that same allele from our mother is also one-half. So a "helping" mutation increases the number of copies of itself by inclining me to care for my offspring or to care for my siblings. The probability that two relatives share an allele is called the coefficient of relationship, denoted r. For two siblings, $r = 0.5$.

So the fitness of the altruistic helper has a direct component (its own survival and reproduction) and an indirect component (the increased fitness of the siblings, or more generally, the related recipients of aid). These two components, together, have been called "inclusive fitness." An altruistic behavior reduces the direct (personal) component and increases the indirect component of the altruist's inclusive fitness, compared with the fitness of a "selfish" genotype.

Whether or not altruism evolves depends on the average fitness benefit (b) to the recipient, the average cost (c) to the individual actor, and the relationship (r) between actor and recipient. The British biologist William Hamilton (one of the most creative evolutionary biologists of the last century) determined a fundamental formula, known now as Hamilton's rule: $rb > c$. This states that altruistic behavior can evolve if the fitness cost to the actor is less than the fitness benefit to the recipient, weighted by r, the coefficient of relationship. The relationship among relatives is highest for parent and offspring ($r = 0.5$) and between siblings (also $r = 0.5$); it is lower between half-siblings that share only one parent ($r = 0.25$), between aunt (or uncle) and niece (or nephew) ($r = 0.25$), and lower still between first cousins ($r = 0.125$). Because the likelihood of shared genes between cousins is less than between siblings, the benefit to the recipient must be greater, relative to the cost incurred by the actor, for altruism to pay off. Natural selection based on aid to relatives is called kin selection. All other things being equal, kin selection is more likely to result in helping offspring, parents, or siblings than anyone else. (And not to help unrelated individuals, with r about zero. Any such instances of helping need another explanation, such as reciprocity or deferred benefit.)

Wild Turkeys (*Meleagris gallopavo*) are a splendid example of kin-selected cooperation. Males gather in large display arenas, where most males form pairs that together court females and defend them against other males—but one male is dominant and typically achieves copulation and fatherhood (plate 31). Alan Krakauer found, from DNA evidence, that most of the dominant–subordinate pairs were brothers, with average $r = 0.42$.[14] On average, dominant males had 6.1 more offspring than solitary males: the benefit (b) of cooperating. Krakauer assumed that the average number of offspring of solitary males (0.9) is the cost (c) of being a subordinate male—the number of his own offspring that he foregoes by helping his brother instead of being solitary. Using these numbers, $rb > c$ is calculated as (0.42×6.1) > 0.9, or 2.56 > 0.9, for a difference of 1.7 offspring: the inclusive fitness advantage of being a cooperative subordinate male rather than a solitary male. Turkeys obey Hamilton's rule. Lance-tailed Manakins and Wild Turkeys both have cooperative male courtship but for very different reasons.

Cooperative courtship is unusual, but cooperative breeding, meaning nesting and brood care, is common among birds. In many cases, young adults, rather than dispersing and immediately reproducing, stay with their parents and help them rear another brood—or broods, for this can last for several years in some species. In some species, helping relatives is a "fallback option" for birds that tried to breed but failed. In general, helpers contribute more to brood care the more closely related they are to the nestlings.[15] Cooperative breeding has evolved independently in at least twenty-eight taxonomic families of birds, and often several times within a single bird family; about 9% of bird species breed cooperatively.[16] There are no simple ecological explanations of why some species are cooperative and others not,[17] but at least one advantage has been identified: protection against cuckoos or other brood parasites. Species that are hosts of brood parasites are more likely to breed cooperatively than nonhost species. In Australia, groups of fairywrens are more effective than single pairs in defending their nest against Horsfield's Bronze Cuckoo (*Chrysococcyx basalis*) and are less likely to be parasitized.[18]

Biologists have done long-term studies on many cooperatively breeding species. I will describe a few case studies that show how complex and sophisticated birds' behavior can be.

Ben Hatchwell, at the University of Sheffield, led a study of Long-tailed Tits (*Aegithalos caudatus*).[19] Pairs start to reproduce early in the season, but about 72% of nests fail because of nest predators. The failed breeders, especially the males, often join another pair and help to rear their young. Both the helped pair and its offspring benefit by having an improved chance of survival to the next year. Not surprisingly, helping has a cost: helpers have a lowered chance of survival. The Hatchwell group found—and this is the key point—that failed breeders strongly prefer to help related adults, whom they recognize as kin by vocal cues that they learned when they were young. Using DNA markers to estimate the relationships (r) between offspring, their parents, and helpers, the researchers estimated r, b, and c. They calculated that Long-tailed Tits pass the Hamilton's rule test: the genetic benefits of helping (rb) exceed the cost (c) by about 67%, mostly because of the nestlings' improved chance of survival. This is an example of kin-selected helping via the fallback option.

Pied Kingfishers (*Ceryle rudis*), common throughout much of Africa and beyond, were the subject of a classic study by Heinz-Ulrich Reyer.[20] Like many kingfishers, they dig a nest burrow in a vertical exposed bank near water. Reyer

FIGURE 9.3. In Pied Kingfishers (*Ceryle rudis*), the number of successfully fledged young is greater the more adults provide care (mostly by feeding). Parents are assisted by primary young males (their older offspring) or secondary, unrelated young males. The two columns at right show that primary helpers make a greater contribution to fledging success. (After Reyer 1990; art, Luci Betti-Nash.)

found that in two kingfisher populations in Kenya, males outnumber females by 1.6 to 1, owing to higher female mortality. (This occurs partly because females spend more time incubating eggs in their burrows, which collapse or are flooded fairly often.) Many one-year-old and two-year-old males can't find a mate. Making the best of a bad situation, they can achieve at least some reproductive fitness indirectly by helping breeding pairs (figure 9.3). Reyer found that pairs have "primary helpers," who are helping their own parents raise some younger siblings, and "secondary helpers," who are unrelated to the pair they help. Primary helpers invest as much effort as their parents do, by guarding the nest and bringing food to both their parents and the nestlings. They have higher inclusive fitness, achieved indirectly by helping kin, than "delayers," who don't help in their first year and who have a moderate chance of getting a mate when they are two years old. Secondary helpers spend much less time and effort helping the breeding pair and increase the nestlings' chance of survival only modestly. But their fitness is greater than a delayer's because "when a male breeder dies, it is usually his former secondary helper who takes over the female." The benefit of cooperation is delayed. Based on his estimates of direct and indirect fitness of males in their first two adult years,

Reyer calculated average inclusive fitness to be 1.76 for the rare male who gets to breed in his first year, 1.09 for a primary helper, 0.92 for a secondary helper, and a dismal 0.30 for a nonhelping delayer. Helping pays off, either via kin selection or delayed benefit.

One of the questions about young birds that help their parents is why they don't leave and start to breed right away. A likely answer, for some species, is that all the space for setting up a territory may be already occupied—the habitat is saturated. Strong evidence for this hypothesis is one of many interesting results of a study of Seychelles Warblers (*Acrocephalus sechellensis*) that Jan Komdeur (professor at the University of Groningen in the Netherlands) started in 1985 and continues to lead.[21] This species lives only in the Seychelles Islands in the Indian Ocean. DNA from museum specimens collected over the last 140 years shows that the population had so much genetic variation among individuals that it must have been quite large—several thousand.[22] But most of the species' habitat was cleared for coconut palm plantations, and by the 1960s, only thirty birds were left. Bird conservation organizations bought one of the islands (Cousin) and restored the habitat. The population has recovered, and some birds have been moved to habitat patches on other islands.

On Cousin Island, most young females, and a few males, help their parents for at least a year, and the survival of the offspring, both in their first year and later in life, is correlated with the number of helpers at a nest. On this island, a territory that becomes vacant because of the death of its owner is immediately filled by a nonbreeding helper. When a few birds were moved to unoccupied habitat on other islands, their offspring dispersed and set up their own territories. In the first few years, none became helpers; only as the new populations grew and filled the habitat did young birds adopt the helping strategy. So helping does seem to be a response to habitat saturation.

(Komdeur and his team have uncovered various complications. The most surprising—in fact, astonishing—discovery was that females can choose the sex of their offspring and that their choice is adaptive. On poor territories [in lower-quality habitat], pairs have higher fitness through raising sons than daughters, and vice versa on good territories [where help by daughters can increase the survival of the nestlings]. Komdeur found that 88% of eggs laid by females on good territories were female, while females on poor territories laid 77% male eggs. This was completely unexpected because a bird's sex is determined by its sex chromosomes: female if ZW, male if ZZ. All sperm carry a Z chromosome, but half of a female's eggs carry Z and half carry W. How a female manages to make an egg more likely to have one or the other is not fully understood, although corticosteroid hormones play a role.[23] I know this is off topic, but it's too interesting to leave out of the story.)

Some birds, apparently, have a remarkable knowledge of their kinship relations. The White-throated Bee-eater (*Merops albicollis*, plate 32), in Kenya, was one of the first, and still among the most interesting, cooperative breeders to be studied in detail—by Steve Emlen at Cornell University. This species lives in large colonies, in which each pair occupies a burrow in a vertical bank that is used for nesting and for roosting in the nonbreeding season. Each individual feeds in a permanent, specific "foraging home range" some distance away from the colony and returns to its burrow before nightfall. The colony is divided into "clans"—extended families that include as many as five mated pairs and some unpaired young or widowed individuals, together extending over three or even four generations. One member—usually the female—of a newly formed pair joins its mate's clan but maintains social ties with its natal clan as well. The clan members forage near one another and visit each other's burrows in the evening. Each bird interacts with many different relatives that may include grandparents, aunts, and cousins—and, Emlen wrote, each "'knows' and behaves differentially toward a large number of specific individuals far beyond its mate and nestlings" (Emlen 1990, p. 504).

About half of nesting pairs have helpers, who are most often older offspring of one or both members of that pair or else failed breeders. Emlen showed that helpers help nestlings who are their younger full siblings or half-siblings much more frequently than would occur if they helped young at random. In 115 cases in which a helper had a choice between more closely and more distantly related nestlings, 94% chose to help the more closely related ones (see figure 9.4). As Emlen wrote, "this ability to discriminate among different classes of potential recipients suggests that bee-eaters possess a fairly sophisticated system of kin recognition."[24] His study was one of the first to show how sophisticated birds can be. He also showed that helping makes a big difference. The rate at which nestlings get fed, and the number that successfully fledge, both increased dramatically with the number of helpers at a nest. Emlen concluded, "By preferentially choosing to aid close, as opposed to distant, kin, a bee-eater can gain almost as much in indirect fitness as a helper as it could in direct fitness as a breeder" (Emlen 1990, p. 524).

Although kin selection may be important in most cooperatively breeding species, some species in at least forty-six taxonomic families breed cooperatively without helping related offspring.[25] There are several possible reasons,

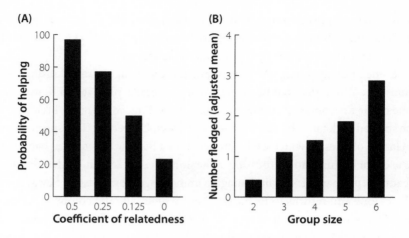

FIGURE 9.4. A long-term study of a colony of White-throated Bee-eaters (*Merops albicollis*; plate 32) established who was related to whom. The left graph shows that the closer the relationship of a bird to a pair's offspring, the more likely it was to help. A coefficient of 0.5 means the helper was the older sibling of the nestlings. The right graph shows that, as in the Pied Kingfisher (figure 9.3), a larger group of helpers is more likely to successfully rear more nestlings to the fledging stage. (After Emlen and Wrege 1991.)

including, simply, that individuals benefit from being in a group, whether of relatives or strangers.

The three species of anis—tropical American members of the cuckoo family—nicely illustrate this theme. Sandra Vehrencamp, who studied Groove-billed and Smooth-billed Anis (*Crotophaga sulcirostris, C. ani*), and Christina Riehl, who more recently studied the Greater Ani (*C. major*), found that these species have fairly similar life histories and behavior.[26] Each nest is typically built by and contains eggs of two or more pairs that collectively defend their nest and territory against other groups. Riehl used DNA markers to establish parentage and other relationships and showed that conesting females are not related to each other. To call these birds cooperative is incongruous because before a female lays her first egg, she ejects any egg that a fellow female has already laid in the nest! Each female stops ejecting once she has laid her first egg. Eventually, all the females have laid at least one egg, and incubation (by males) begins. Despite this conflict, communal nesting is highly advantageous because larger groups suffer less nest predation. Three-pair Greater Ani nests were much more successful than two-pair nests, and the number of successful fledglings per female was correspondingly greater. (Single pairs almost never nest alone,

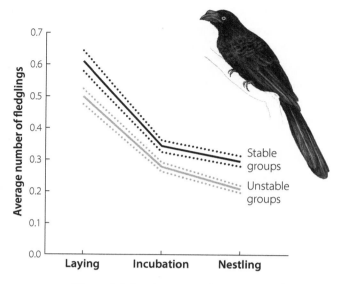

FIGURE 9.5. The probability that a nest of Greater Anis (*Crotophaga major*) with eggs or nestlings will survive through three stages of nesting. The upper line is for nests of stable groups of females that had already cooperated the previous year. The lower line shows a lower success rate for groups that had lost a member or added a new member since the previous year. (The dotted lines are a measure of the statistical uncertainty about the actual values.) (From Riehl and Strong 2018; art, Luci Betti-Nash.)

and the few that Riehl has seen failed.) What is more, females that have previously bred together synchronize their egg laying more rapidly, which reduces competition and increases their reproductive success because it reduces the length of time the eggs and nestlings are at risk of predation by snakes, monkeys, and caracaras (figure 9.5). So we have come full circle, to cooperation based entirely on self-interest. Occasionally, nests are destroyed, and in that case, the females abandon any semblance of cooperation and become brood parasites, laying their eggs in other groups' nests. Riehl found that parasitic females lay more eggs than host females (more than compensating for their loss) and that parasitic females and host females have about the same level of reproductive fitness.

What do these various species tell us about how and why social behaviors have evolved in birds? That certain behaviors enhance the individual's personal, direct fitness, that some improve fitness indirectly by benefiting the individual's relatives, and that some can have both effects. Joining a group can be a "selfish" act, as when it provides protection from predators or greater future chances of mating. Or it can be "altruistic," as when cooperative breeders aid their parents or other relatives to raise more offspring. In almost all cases, social interactions, however cooperative they may be, still include real or potential conflicts of interest, just as in human societies. Still, what has emerged, in some mammals and many birds, are elaborate, intricate societies that depend on advanced intelligence or cognitive ability: knowing other individuals personally, remembering at least some past interactions, modulating one's behavior to the ever-changing social context. The natural selection that has shaped many aspects of bird behavior has arisen from individuals' social environment more than the external, ecological environment that we may have learned is the source of natural selection. The behavior of bird species shapes much of their own evolution.

10

Bird Species

WHAT ARE THEY AND HOW DO THEY FORM?

Almost all my birding friends (and I) keep lists: how many species they have seen in their life, how many this year, how many in their yard, their favorite "patch," their county, state, country, the world. One of my friends lists species seen through his car's sunroof. The number of species recorded in a county or state is generally straightforward—but at the level of country, continent, or the world, it can be ambiguous, and the birder may be faced with decisions. How many of the birds I have seen are distinct species?

The number of "recognized" species changes all the time and differs among books and among "official" lists. As of October 2019, the number of species in the world was 10,721 in the Clements Checklist; 10,758 in the IOC World Bird List; and 10,021 in the Howard and Moore Checklist of the Birds of the World.[1] But one research team, extrapolating from variation among specimens of 200 species, has provocatively proposed that there are about 18,000 species of birds in the world![2] What's going on? What are species?

New species names are added to these lists in three ways. The most exciting is the discovery of species that have never been recognized, or even seen before, by scientists. In *Birds New to Science: 50 Years of Avian Discoveries*, David Brewer portrays the discovery of 288 species since 1960.[3] The spectacular Okinawa Rail (*Hypotaenidia okinawae*), a large rail with a bright red bill, was first seen and subsequently collected by one of the ornithologists who named it (plate 33). The Long-whiskered Owlet (*Xenoglaux loweryi*), described as "utterly unlike" any other owl, was found in a mist net in an ornithologically little-known area in Peru. (A lodge near this site now caters to the many birders—including me—who come to see the species.)

Some new species are recognized when an ornithologist undertakes a detailed study of some well-known species and discovers that a "single species" includes two or more look-alikes that differ in other features, such as their songs. The bird I learned as "Traill's Flycatcher" was formally recognized in 1973 to include two fully distinct species—now called Willow and Alder Flycatchers (*Empidonax traillii, E. alnorum*)—that have quite different songs and often nest in the same swampy area without interbreeding. Similarly, vocal differences are among those that distinguish the four new species of screech owls, eight new scops owls, and seven new pygmy owls listed in Brewer's book.

The most common way "new species" are recognized is by elevating the status of different geographic populations of a species that are already known to have distinctive features: "splitting" one species into two or more. Often, these distinctive populations had already been named as subspecies, the taxonomic term for geographic "races." The 2019 update of the Clements Checklist includes 106 newly recognized species based on "splits," although it lost 20 owing to "lumping" them with other species. North American birders lamented the loss of "Thayer's Gull" in 2017, when its long-expected union with Iceland Gull (*Larus glaucoides*) became official, but they may have welcomed several recent splits. The eastern and western populations of the Winter Wren in North America were formerly subspecies (*Troglodytes troglodytes hiemalis* and *T. t. pacificus*, respectively) and various Eurasian populations of *T. t. troglodytes* were considered the same species as the Winter Wren. Research on their vocalizations and genetics led to a three-way split into the Pacific Wren (*T. pacificus*) and the Winter Wren (*T. hiemalis*) in North America and the Eurasian Wren (*T. troglodytes*).[4]

Almost all the cases in which there is a difference of opinion about the number of species have to do with splitting versus lumping: should we recognize distinguishable groups of birds as one species or more than one? There are two distinct issues: the philosophical question "what is a species?" and the practical question of whether a particular population or lineage is a species (by whatever philosophical concept of species we adopt). There never has been and perhaps never will be full agreement among taxonomists and evolutionary biologists on either of these points. Some of the difficulties arise from very different opinions about what "species" *should* mean, and some arise from the nature of evolution and the processes by which species (by any definition) evolve.

What are species? Maybe it's best to start with what we now think is the process of speciation and then look at different species concepts in its light.

Speciation is the origin of two (or possibly more) species from a single ancestral species. The usual process involves geographically distinct (*allopatric*) populations of the ancestor that form as the species spreads into new areas or when its distribution becomes interrupted by, say, change in the course of a river. Imagine two such areas separated by water or by terrain that the species doesn't inhabit and seldom crosses, so that there is little movement of individuals—and little exchange of genes—between the populations. Evolutionary changes by mutation, natural selection, and genetic drift (chapters 4 and 5) will then transpire independently in the two populations, which become more and more different over time.

Three somewhat overlapping kinds of genetic differences may evolve. Some have little or no effect on fitness (survival or reproduction) and evolve by genetic drift. Many DNA sequence differences, especially synonymous base pair changes (those that do not change an encoded protein), and possibly some slight differences in plumage or other characteristics are in this category. Second, some phenotypic features that affect survival and reproduction diverge by natural selection owing to different ecological environments or by social or sexual selection. Third, some of these changes may lead to *reproductive isolation* of the two populations, should they ever encounter each other and become *sympatric* (occupying the same area). Bird populations may mate mostly or exclusively with their own kind, or they might produce hybrid offspring that have reduced survival or reproductive success. These are ways in which the exchange of genes between the diverging populations can be curtailed even when there is opportunity to interbreed. All of these kinds of differences are likely to evolve gradually and to become more pronounced as time passes. So no matter what species concept is used, it will be impossible to specify a point at which the populations suddenly qualify as species. (Just as there is no objective basis for deciding exactly when a youth becomes an adult.)

The main contending concepts of species used by ornithologists today are the "biological species concept" (BSC) and several variants of a "phylogenetic species concept" (PSC). Both concepts view species as more or less independently evolving lineages. The PSC considers populations to be distinct species earlier in their history of divergence, when they differ in at least one gene or character, than does the BSC, which requires that divergence has progressed to reproductive isolation.[5] The BSC emphasizes the current status of populations, their degree of interbreeding at this time. The PSC is a more historical concept, emphasizing the division of an ancestor into different descendants.

From the days of Linnaeus, in the mid-eighteenth century, to the early twentieth century, species were distinguished and classified almost entirely by discrete differences in morphological features of museum specimens. As taxonomists studied more and more specimens, especially over broad geographic regions, they confronted variation that challenged their ability to classify them. Morphologically different forms (such as the color morphs of the Ruff) are sometimes the same species. Differences between birds from different regions ranged from slight to great. Morphologically almost identical birds sometimes proved to include two or more distinct forms.[6] Among the first cases was a study, in 1919, of the Treecreeper by Erwin Stresemann, Germany's leading ornithologist at the time, who recognized two species (*Certhia familiaris, C. brachydactyla*) with overlapping geographic distributions, based on subtle but consistently associated differences in plumage and claw length.[7]

Stresemann's star student was Ernst Mayr, who made abundant contributions to bird taxonomy but is most widely known for his great influence on evolutionary biology. In *Systematics and the Origin of Species* (1942), Mayr articulated the "biological species concept," that emphasized the theme of reproductive community or disjunction: "Species are groups of actually or potentially interbreeding natural populations, which are reproductively isolated from other such groups."[8] The BSC has been the framework of most subsequent research on how new species evolve from their common ancestors.

The key words in Mayr's definition are "reproductively isolated" and "actually or potentially interbreeding": two sides of the same coin. Mayr and later researchers cast these ideas in terms of gene exchange, or "gene flow." Think of genes descending, by DNA replication, from generation to generation. A Song Sparrow (*Melospiza melodia*) born this year in Nova Scotia might have some genes descended from ancestors in Massachusetts, if over the course of generations, some descendants of the Massachusetts birds dispersed northward and interbred with the residents. This would be gene flow, via a chain of *actually* interbreeding local populations from Massachusetts to Nova Scotia. But Song Sparrows might disperse very seldom from the Nova Scotia mainland to Cape Sable Island, 175 kilometers away. Nonetheless, they almost surely are *potentially* able to interbreed—to mate and have surviving, fertile offspring, if some birds do fly across the strait.

What does "reproductively isolated" mean? Any biological difference between populations that reduces gene flow between them. I often meet people who were taught that organisms are different species if crosses between them result in sterile offspring. (The mule is the standard example.) This is indeed one form of reproductive isolation, a form of "postmating isolation," which occurs if mating takes place but the hybrid offspring fail to develop, or are likely to die before they reproduce, or have low reproductive success because of sterility or inability to get mates. However, closely related bird species often have the potential to produce fertile hybrids but simply don't mate with each other; they have "premating" reproductive isolation.

In some cases, premating reproductive isolation occurs because a biological difference has evolved between the populations that prevents them from meeting. The Band-rumped Storm Petrel (*Oceanodroma castro* complex) has populations that breed on the same islands but in different seasons. They have some DNA sequence differences and are now considered separate species.[9] But by far the most important kind of reproductive isolation between closely related bird species is "behavioral isolation"; females mate preferentially or exclusively with males of their own species, based on plumage, display behavior, song, or some combination.[10] Myron and Ann Baker provided female Indigo Buntings and Lazuli Buntings (*Passerina cyanea, P. amoena*), closely related species that sometimes hybridize in nature, with separately caged males of the same or the other species and at the same time exposed them to recordings of the conspecific (same species) or allospecific (other species) song.[11] The females solicited copulation more often from the combination of conspecific male and conspecific song than from conspecific males with allospecific song, or vice versa. (And they had no interest at all in the allospecific male + song combination.) The authors concluded that females respond both to the male appearance (probably plumage) and song. In another experiment, Laurene Ratcliffe and Peter Grant exposed wild Darwin's finches to stuffed birds of the same or other species by use of different combinations of detached heads and bodies.[12] Wild birds of both sexes reacted aggressively, and males sometimes courted female specimens. Overall, the birds responded most to models that had both the head and body of their own species, suggesting that they recognize their own species based on both plumage and beak differences. In some cases, the stronger response to a conspecific cue may result from learning; most oscine songbirds seem to have an inherited "template" of the basic form of their song, but they learn the particulars from their father or other nearby males.[13] Still, there is usually some genetic component to both the male signal (plumage, song, display) and the female reaction, which must evolve more or less together.[14]

In practice, biologists determine that sympatric species are reproductively isolated not by such experiments but by inferring isolation from the pattern of morphological features or genetic markers. (The morphology doesn't *define* species; it is used as *evidence* of the species criterion, reproductive isolation.) We infer that birds are different species if we find few or no hybrids between them. Hybrids can be recognized by their intermediate features or a mixture of features that distinguish the species. For example, Blue-winged Warblers (BWWA, *Vermivora cyanoptera*) and Golden-winged Warblers (GWWA, *V. chrysoptera*) have hybridized as the BWWA has extended its range (figure 10.1). The GWWA has a black throat and ear patch (lacking in BWWA), a white belly (yellow in BWWA), and yellow wing bars (white in BWWA). First-generation (F_1) hybrids, known as "Brewster's Warbler," resemble BWWA in lacking the black throat and ear patch, and they are intermediate in the color of the belly and wing bars. The rarely seen "Lawrence's Warbler," the backcross offspring of "Brewster's" × GWWA, combines GWWA's black throat and ear and BWWA's yellow belly and white wing bars.[15]

Whether or not populations hybridize can also be revealed by genetic markers such as SNPs (single nucleotide polymorphisms, chapter 5): DNA sites where different nucleotides (say, A and G) are found in different gene copies. If the frequency of a nucleotide differs substantially between two groups (say, 90% A in one, but only 15% in the other), and if a similarly strong difference is found at many SNPs in different genes, we may suspect that there is little gene exchange, which would homogenize the populations. If the populations have very similar frequencies, we infer that they have a history of extensive gene flow between them. Alder and Willow Flycatchers, which are vocally distinct but morphologically almost identical, differ consistently in their mitochondrial DNA sequences: they evidently do not hybridize. In contrast, Cordilleran and Pacific-slope Flycatchers (*Empidonax occidentalis*, *E. difficilis*), formerly considered a single species (Western Flycatcher), have mostly allopatric distributions but meet in southwestern Canada. Some genetic markers differ in frequency between the two forms, although none are completely different. Most of the birds sampled in the contact zone could not be assigned to either "species" but instead showed mixed ancestry.[16]

As you read about the hybridizing warblers, you might have thought, "Hold on! If those warblers hybridize, and the "Brewster's" hybrid can cross with the parental types, why are they species? Doesn't that contradict the definition?" In practice, no evolutionary biologists—not even Ernst Mayr—have ever insisted that species be 100% reproductively isolated. DNA-based studies have shown that many, maybe most, very closely related species hybridize occasionally and that at least some genes make their way from one to the other. In a few

Blue-winged Warbler   Golden-winged Warbler

"Brewster's" Warbler   "Lawrence's" Warbler

FIGURE 10.1. Blue-winged (*Vermivora cyanoptera*) and Golden-winged (*V. chrysoptera*) Warblers, and their hybrids: "Brewster's" and "Lawrence's" Warblers. (Art, Luci Betti-Nash.)

cases, many genes cross the species border. In a recent study of the Blue-winged and Golden-winged Warblers, David Toews and colleagues found that their genomes are indistinguishable except at six small genomic regions, most of which include genes known to affect coloration, including the color pattern differences between these birds.[17] Nevertheless, the "Brewster's" and "Lawrence's" hybrids are rarely seen, and Frank Gill, who studied these birds in depth, wrote that "the frequency of hybridization . . . is not high enough . . . to warrant a change in species status."[18] There is some evidence that sexual selection may maintain the difference in plumage. Researchers used bleach to erase the black throat and mask of some territorial male Golden-winged Warblers so that they resembled the "Brewster's" hybrid.[19] Compared with control males, more of these birds were evicted by other males and lost their territories. If hybrids suffer the same fate, they would have reduced reproductive success—hence, some degree of reproductive isolation between the species.

The BSC has been the preferred species concept of most evolutionary biologists because they ascribe special significance to reproductive isolation: it helps to maintain species as distinct, independently evolving lineages that may become more and more different in the future. (Every branching point in a phylogenetic tree marks a past speciation event.) But even aside from borderline cases like these warblers, putting the BSC into practice presents a major difficulty: how can we tell if allopatric populations, which occupy distinct geographic regions, are "potentially interbreeding" or would be reproductively isolated if they came into contact? Geographically separated populations show all degrees of difference, from slight to great. The "pied monarch complex" on various Melanesian islands east of New Guinea is one of many examples (figure 10.2). Near the end of his life, Ernst Mayr, the world authority on Melanesian birds, resorted to calling many of these forms "allospecies," having no way of telling whether or not they are potentially reproductively isolated from each other.[20]

What to do? We usually can't do experiments, as the Bakers did with buntings, and it is ethically unacceptable to move wild birds around to see if they interbreed. In practice, ornithologists have used sympatric species as a guideline: call the allopatric forms different species if they are about as different (especially in plumage, vocalizations, or both) as sympatric, reproductively isolated members of that group of birds (in, say, the same genus).[21] A committee of the British Ornithological Union proposed that allopatric taxa be assigned species rank if they can be completely distinguished by each of several characteristics and the sum of the character differences corresponds to the level of divergence in related pairs of coexisting sympatric species.

Advocates of the phylogenetic species concept (PSC) would name allopatric populations as species if they can be reliably distinguished ("diagnosed") by any feature whatever.[22] To cite one of many examples, Robert Zink studied variation among geographical populations of the Fox Sparrow, then considered a single species (*Passerella iliaca*) with four major geographic groups that differ mostly in plumage color (plate 34). He found that the four groups differ in mitochondrial DNA, which also indicated some interbreeding between the groups where their ranges meet.[23] Both the IOC and the Clements world checklists have adopted Zink's proposal that these groups be recognized as phylogenetic species, even though they interbreed to some extent. Zink, a champion of the PSC, was one of the authors of the study that I mentioned earlier, suggesting that the number of bird species might well be twice the currently recognized number. Frank Gill, who has led the creation and ongoing revision of the IOC world checklist, recognizes reproductive isolation as an important criterion of species, but he argues that morphologically and vocally different allopatric populations are much more likely to be potentially

FIGURE 10.2. Four of the many different forms of pied monarchs (genus *Symposiachrus*) that inhabit different islands in Melanesia. Although difference in color pattern might affect mating, experiments would be needed to tell whether or not they are species in the sense of the biological species concept. Several of these are now recognized as species, under a phylogenetic species concept. (Art, Stephen Nash, based in part on Mayr and Diamond 2001, courtesy of H. Douglas Pratt).

reproductively isolated than we have generally thought. So he proposes that these populations be named distinct species unless there is good reason to think otherwise.[24] I'm an advocate for the traditional BSC, but I admit he makes a fairly persuasive argument. (However, I take a dim view of a trend toward calling populations different species just because they have some DNA differences. Many such differences are neutral mutations that have no effect on the birds' features or biology; they simply tell you that the populations have been geographically more or less separated.)

The choice of a species definition is largely a matter of convention, but how new species evolve is a scientific question.[25] If we adhere to the BSC, we would focus our attention on explaining the features that cause reproductive isolation between the species. In fact, a supporter of the BSC (such as I) would *define* speciation as the evolution of reproductive isolation between descendants of a common ancestral species.[26]

Ernst Mayr compiled voluminous evidence that speciation in birds and other animals almost always depends on division of the ancestral species into spatially separated populations that evolve independently. This is geographic, or allopatric, speciation. In contrast, sympatric speciation means that one species splits into two without any initial spatial separation of the diverging populations. There are a few plausible examples. I mentioned the Band-rumped Storm Petrel, in which populations on the same islands have evolved disjunct breeding seasons.[27] Another possible case is in several closely related species of African indigobirds, each of which is a brood parasite on a different species of finch. Both sexes become "imprinted" on the song of the male foster parent: males mimic it and females respond to it. It is plausible that an occasional shift to a new, sympatric host species with a different song could initiate a new species of indigobird.[28]

However, geographic speciation is the general rule in birds. Some of the strongest evidence comes from birds on islands. Large islands, such as Madagascar, harbor pairs of sister species (closest relatives), the result of in situ speciation. But Jerry Coyne and Trevor Price could find no evidence of sister species on any small, isolated oceanic islands, which evidently do not provide the spatial separation required for divergence.[29] Darwin's finches have speciated plentifully in the Galápagos archipelago, where populations can diverge on separate islands and later disperse among islands and coexist. Long ago, one species of this group colonized remote, isolated Cocos Island, but it has remained a single species, even though individuals specialize in different ecological niches (see chapter 6).

Many hybrid zones also give evidence for allopatric speciation. A hybrid zone is a region where genetically different populations in different areas meet and interbreed. If the first-generation hybrids are at all fertile, they may back-cross with parental types, so genes from each population may spread (or "introgress") in subsequent generations into the other population. The lower the fitness of the hybrids, the lower the rate of spread of genes between the populations. But in some cases, certain genes or characteristics are more disadvantageous than others in the receiving population. This explains why some genes and characteristics sometimes introgress further than others, so that the width of the hybrid zone differs for different parts of the genome.

The Red-legged Partridge (*Alectoris rufa*), in Spain and southern France, hybridizes in the Alps with the eastern Mediterranean Rock Partridge (*Alectoris graeca*).[30] The species are thought to be about 2 million years old, based on a "molecular clock," but they came into contact in the Alps only 6,000 to 8,000 years ago, when the glaciers melted back. Birds with various hybrid combinations of plumage features occur in a zone about 15 kilometers wide, but some genes have trickled from each species at least 90 kilometers into the range of the other species (figure 10.3). Using rough estimates of how far these birds disperse, per generation, Ettore Randi and Ariane Bernard-Laurent calculated that genes should have moved between the species over a much greater distance by now (1,120 to 2,750 kilometers). Some kind of natural selection must be impeding their spread. These researchers found that birds with certain combinations of Red-legged and Rock Partridge genes were less common than would be expected. They propose that many mixed genotypes have low fitness because of genetic incompatibility. There is enough postmating isolation to keep the species more or less separate, even though they hybridize.

Another example, which has been studied since the 1920s, is the western Carrion Crow (*Corvus corone*, all black) and the eastern Hooded Crow (*C. cornix*, gray with black head, wings, and tail), which form a hybrid zone 10 to 100 kilometers wide, from Denmark to northern Italy. Like the partridges, these crows are thought to have met after the retreat of the European ice sheet, spreading from "refuges" in Spain and the Balkans. Hybrid females have fewer surviving offspring than nonhybrids, and birds of both species are more likely to mate with their own species than with the other. In two recent studies, researchers found that many genes have spread between the species, far beyond the zone that was recognized by plumage variation. Only three short regions in the genome differed strongly between the species. These regions all contain genes that control the synthesis and pattern of melanin

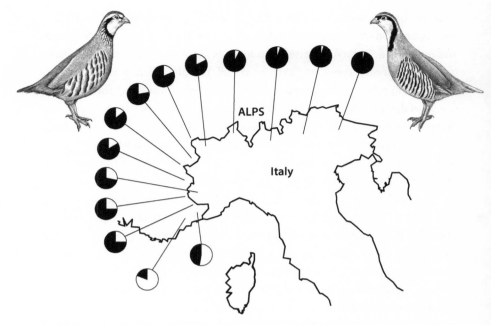

FIGURE 10.3. A hybrid zone between the eastern Red-legged Partridge (*Alectoris rufa*) and the western Rock Partridge (*A. graeca*). Each pie diagram shows, for a sample of birds in a certain location, the average proportion of genetic markers (single-nucleotide polymorphisms) that are characteristic of the Red-legged Partridge (in black) or the Rock Partridge (in white). The genetic composition changes gradually across the hybrid zone from east to west. (After Randi and Bernard-Laurent 1999; art, Stephen Nash.)

pigments. It appears that the birds rely on coloration for mating decisions, imposing strong selection against the incursion of the "wrong" coloration genes into each species' range.[31]

The most challenging question we can ask about speciation is: why does it happen? What causes different populations of a species to evolve reproductive isolation?

The causes of postmating isolation, meaning reduced survival or reproduction of hybrids, can be extrinsic (due to environment) or intrinsic (often due to incompatibility between genes). Peter and Rosemary Grant found that hybrids between two species of Darwin's finches had very low survival in dry years, when the available seeds were mostly large, but excellent survival in wet years, when small seeds predominated.[32] But low fitness of hybrids often results from genetic incompatibility, which can cause sterility and other de-fects. Wild-caught hybrids between two species of chickadees were less profi-cient than nonhybrid birds at learning how to get food from behind a simple barrier. Captive-reared hybrids between two species of small parrots (love-birds) made a variety of mistakes in cutting and transporting nest material.[33] So far, there doesn't seem to be any detailed evidence about how specific genes interact to cause incompatibility in hybrids, and why the genes diverged. But genetic incompatibility is seldom the cause of speciation in birds simply because premating isolation almost always evolves long before incompatibility does. Hybrids seem to be fully fit in about half of 254 different crosses between closely related species, many of which diverged as long as 5 Ma.[34] We can learn more about the causes of speciation if we focus on premating isolation, espe-cially behavior. What causes behavioral isolation to evolve?

"Ecological speciation" occurs when reproductive isolation arises as a result of divergent natural selection and adaptation to different environments or re-sources.[35] One possibility is divergence by "sensory bias," which I mentioned in chapter 8: the idea that mating or social signals evolve to be conspicuous in the species' environment. For instance, it has long been known that the songs of open-country birds tend to include higher frequencies than those of birds that live in forests, where foliage damps these frequencies. Joseph Tobias and colleagues compared the songs of pairs of closely related species of antbirds, furnariid ovenbirds, woodpeckers, and tinamous that occupy typical Peruvian rainforest versus stands of bamboo.[36] They showed that high frequencies are attenuated more in bamboo and that the songs of bamboo-living species have lower frequencies than those of their relatives.

Reproductive isolation could also be a "side effect" of ecological adaptation. I mentioned earlier that both sexes of Darwin's finches seem to recognize con-specific mounted birds partly by their bill size. Furthermore, bill size affects the rate at which a bird can open and close its mouth—and thus the frequency range and the rate of successive notes. So the divergent adaptation of finch bills to different diets resulted in divergence of signals used in mating.[37] Elizabeth Derryberry and colleagues described a similar association between bill evolu-tion and song divergence in Neotropical woodcreepers. And the Red Crossbill is a poster bird, par excellence, for ecological speciation. Crossbills' bills are highly specialized for extracting seeds from conifer cones. In North America there are different "call types," at least six of which specialize on different co-nifers and have bill differences adapted for the different cones (see figure 5.2).

The calls create flock cohesion: birds of the same call type forage together and choose mates from within the flock. The call types seem to be newly forming species. One of them is adapted to lodgepole pines that have distinctive cones in a region that lacks squirrels. (Pines evolve cone features to deter squirrels from eating seeds.) Based on distinctive genetic markers, Craig Benkman and colleagues showed that this crossbill does not interbreed with other call types. They named it a distinct species: Cassia Crossbill, *Loxia sinesciuris*—meaning "without squirrels."[38]

Another reason behavioral isolation might evolve is to prevent disadvantageous hybridization. Suppose two allopatric populations of an ancestral species have become quite genetically different and now expand their range and come into contact, forming a hybrid zone. It might be advantageous for females to discriminate against males of the other population, if the hybrid offspring have low fitness. (Meaning that some degree of hybrid sterility or other postmating isolation has evolved.) Hybrid offspring of a nondiscriminating female will fail to pass on her genes, whereas any mutations that incline females to mate only with their own kind of male will be inherited by reproductively fit nonhybrid offspring and will therefore increase from generation to generation. The postmating isolation would then be reinforced by the evolution of premating isolation.

This process has occurred in the European Pied Flycatcher, which hybridizes with its closest relative, the very similar Collared Flycatcher.[39] The hybrids have low fitness: hybrid males are less successful in mating and female hybrids are sterile. For these reasons, individuals that do not hybridize would have higher fitness. Outside the hybrid zone, Pied males, true to their name, are black and white and closely resemble Collared males. But in the hybrid zone, most are brown and white, and females prefer these males (plate 35).

The most tempting explanation of how populations come to be behaviorally isolated must surely be sexual selection: as I described in chapter 8, it accounts for many male display features and for female preferences for them (or vice versa). Several research groups have used DNA-based phylogenies to ask if the proliferation of species is greater in bird groups that are thought to experience stronger or more variable sexual selection. There isn't a definite answer yet, but it seems likely to be "yes." Many of the leading researchers in this field collaborated in a comprehensive analysis and found that among very recent speciation events in passerine birds, male plumage traits that are probably sexually selected (such as color) diverged more than female plumage traits or traits related to locomotion and foraging. The divergence in male plumage (and presumably female mate choice) allows the populations to over-lap and stay separate without fusing.[40]

Is there any direct evidence that allopatric populations of a species become different because of sexual selection? And why would sexual selection come to differ between populations? Rebecca Safran's research team found that female Barn Swallows (*Hirundo rustica*) in northern Europe prefer males with longer tail streamers (outermost tail feathers); in North America, however, where Barn Swallows have shorter tail streamers and darker, reddish bellies, females select mates by belly color, not streamer length. In the eastern Mediterranean region, sexual selection favors both long streamers and dark bellies. After a heroic amount of research, the team showed that these features indicate the males' resistance to certain parasites. In the Czech Republic, males with long streamers have fewer lice and nest mites and consequently have more surviving offspring. (And their belly color is not correlated with parasite load.) In Colorado, males with a redder belly have fewer nest mites; and in Israel, streamers signal resistance to lice and throat color resistance to nest mites.[41]

Male Common Yellowthroats (*Geothlypis trichas*) have a bright yellow throat, of course, and a dramatic black mask (plate 36). Linda Whittingham and her colleagues studied the major histocompatibility complex (MHC), which is part of the innate immune system.[42] The many genes in the complex all have various alleles that encode slightly different proteins, which are thought to defend against different pathogens. The research team showed that in both Wisconsin and New York, males with larger black masks are better in male-male competition. Females in the Wisconsin population prefer males with large masks, but in New York, they respond only to the size and bright-ness of the male's yellow throat. Whittingham's group found that Wisconsin males with larger masks have more MHC alleles, fewer blood parasites, and better overwinter survival. In the New York population, the diversity of a male's MHC alleles wasn't strongly correlated with mask size, but it was cor-related with the brightness of his throat. So in both populations, females favor the plumage ornament that is more indicative of a male's MHC variation—which might mean that their offspring could inherit beneficial MHC alleles and be more resistant to diverse pathogens.

We will need more studies before ornithologists will know how often the several sources of ecological and sexual selection cause new bird species to evolve. But there are more questions to ask. How fast is speciation? How long does it take? Can it go into reverse? Are the largest bird families those in which speciation is fastest?

Some of these questions can be answered using a "molecular clock," especially together with phylogenies. Trevor Price and Michelle Bouvier compiled information on both wild and captive crosses and found that 157 out of 254 crosses between species in the same genus produced fully fertile hybrids. It takes several million years of separate evolution for substantial infertility to develop.[43] But many pairs of species are much younger, having evolved behavioral isolation. For example, allopatric sister populations of Neotropical suboscine birds (such as flycatchers) discriminate against each other's songs after about 1 million years of separation. (But it takes oscines, such as wrens, almost four times as long. The difference is that oscines learn some aspects of their song but suboscines do not.[44])

Hybrid zones also provide information. The mitochondrial DNA difference between pairs of hybridizing species in northern latitudes suggests that they evolved separately in allopatry for about 1.5 million years before they met and started hybridizing. So it must take somewhat longer for full reproductive isolation to evolve. Curiously, tropical hybridizing pairs are older—about 3 or 4 million years. Jason Weir and Trevor Price suggested that the speed of speciation has been higher in northern birds because the climatic disruption of northern ecosystems during the Pleistocene glaciations opened up opportunities for birds to adapt to different resources, and reproductive isolation (ecological speciation) was a side effect.[45] If they are right, the faster speciation rate is a result not of something special about today's cold or seasonal environments but simply historical circumstance.

What happens when allopatric populations diverge and then expand their range and meet? In some cases, there is hybridization, as we have seen, and this can have diverse results. Very rarely, a third species may evolve from the hybrid population. The best-known example is the Italian Sparrow (*Passer italiae*), which originated by hybridization between the widespread House Sparrow and the Mediterranean Spanish Sparrow and is partly reproductively isolated from both of those species where they meet (in northern and southernmost Italy, respectively). The Golden-crowned Manakin (*Lepidothrix vilasboasi*), in Amazonian Brazil, is also the result of hybridization between two species. Its unique crown coloration is thought to result from a mixture of those species' genes, together with natural selection for other genes that accentuate the resulting color.[46]

In at least a few cases, hybrids have high fitness. The Glaucous-winged Gull (*Larus glaucescens*) and the more southern Western Gull (*L. occidentalis*) form a hybrid zone about 550 kilometers long on the northwestern coast of North America. Thomas Good and collaborators traced the reproductive success of pairs in two colonies, near both ends of the hybrid zone.[47] In both cases, pairs with hybrid males fledged more offspring than pairs with a "pure" male of either species. Compared with Western males, hybrids held territories in safer, less exposed habitats, and compared with Glaucous-winged males, they foraged more for fish, a richer food that enhanced chick growth and survival. In such cases, we can suppose that progress toward the evolution of fully separate species might simply stop.

Hybridization may set back or undo speciation for other reasons, as well. In many cases, gene exchange is so high that it reverses previous genetic divergence, and the populations fuse. Rosemary and Peter Grant have observed such fission and fusion in Darwin's finches, and one of the three species of tree finches on one of the islands has almost disappeared since the nineteenth century because of hybridization with the other two.[48] This must often be the case for populations that rejoin after only a short period of geographic isolation.[49] I suspect that such "ephemeral speciation" will be the fate of many species recognized under the phylogenetic species concept.

The fate of incipient or newly formed species can also depend on ecological competition. In many cases, they may overlap to some extent without interbreeding and simply coexist by differing in habitat or food. In other cases, the new species may fail to interpenetrate each other's range because of competition. Among tropical furnariid ovenbirds, it takes an average of 11.9 million years for sister taxa to become sympatric, but those that differ in body size or bill shape become sympatric much faster: they clearly have different feeding niches and presumably compete less.[50]

Whether or not newly formed species can coexist may be important to the rate of speciation, which is the average number of new species that arise over a long time span. For example, according to a time-calibrated DNA-based phylogeny (figure 10.4), an ancestral species of wood warbler split about 6 Ma into two species; one became the ancestor of the Golden-winged, Blue-winged, and four other species of warblers, and the other gave rise to the two species of waterthrushes. The rate of speciation, then, was six versus two species per 6 million years, or a threefold difference. What is curious is that in many groups of birds and other animals, the long-term *rate* of speciation is not correlated with the *speed* of speciation: how long it takes for substantial reproductive isolation to evolve between two allopatric populations of an

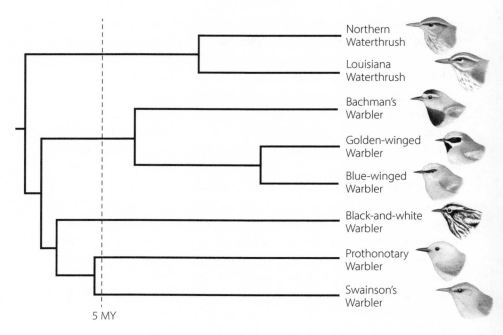

FIGURE 10.4. A branch of the phylogeny of the warbler family Parulidae, showing two clades that diverged from a common ancestor about 6 million years (MY) ago. One clade, the water-thrushes, has one-third the number of species of the other clade, implying a threefold difference in speciation rate between the clades. (After Lovette et al. 2010; Art, Luci Betti-Nash.)

ancestral species. Some authors, then, propose that the long-term rate depends not only on reproductive isolation but also on other factors—perhaps how ecologically different new species become. When species compete less and can coexist, they can expand their ranges. A larger range means more opportunity for local populations of each new species to again become quasi-isolated, experience different environments, and repeat the speciation process. So the rate of evolution of species diversity may depend on the evolution of ecological differences.[51]

It is possible, even likely, that special features or peculiarities of certain groups of birds predispose them to rapid or slow speciation rates. Song is an important way that many species use to choose mates (often together with plumage and display behavior). In fact, a phylogenetic study of the families Thraupidae (tangers and relatives) and Furnariidae (ovenbirds, woodcreepers, and relatives) revealed that there have been bursts of speciation in both groups (when many species arose within a short time) and that these were correlated with bursts of change in vocal characteristics (such as peak frequency and song length).[52]

Does the rate of speciation determine why some families of birds have more species than others? This brings me to a sweeping topic, the world of bird diversity in all its thrilling glory.

11

A World of Birds

Its name alone—Shoebill in English, *Balaeniceps rex* (king whale-head) in Latin—tells you that it's an odd bird, in fact so odd that it is in its own family. A lengthy, close view of one from a small boat on the Victoria Nile River in Uganda was among my most anticipated and most rewarding birding experiences (plate 37). Uganda is definitely the best place to see one. While you are there, you can enjoy a host of other families that are endemic—restricted—to Africa: turacos, mousebirds, honeyguides, ground hornbill, ostrich, and more. If, instead, you go to Australia, you will enjoy cockatoos, fairywrens, pardalotes, gerygones, and, if you are lucky, a cassowary or Plains-wanderer (*Pedionomus torquatus*), the single member of its family. Australia also has a few birds-of-paradise, but their great and astonishing variety is in New Guinea, along with specialties such as jewel-babblers and pitohuis, the world's only poisonous birds. Birders travel to southern oceans to see diverse albatrosses and petrels, to Europe and northern Asia for tits, fringillid finches, buntings, and pheasants, or to North America for prairie chickens, thrashers, and stunning wood warblers. For endless, amazing diversity, go to South America, the "bird continent," for its tinamous, trumpeters, potoos, toucans, puffbirds, manakins, antbirds, spinetails, almost 300 species of hummingbirds, and about 360 species in the tanager family.[1]

It is no wonder that "world birding" has become popular. Some birders focus on building their life list into the thousands of species; some want to photograph them; many just want to experience the variety, beauty, and novelty of the world's birds. Traveling for birds has other rewards as well: you may enjoy different cultures, and you may help conservation by contributing to the economy of local people who become motivated to protect endangered species (plate 38).

That each group of birds has a certain geographic distribution is a familiar fact that actually calls for explanation. In Darwin's time, the answer to a question such as, "Why are sunbirds only in the Old World, but hummingbirds only in the New World?" would usually have been to the effect that it pleased God to put them there. Darwin recognized that if species have evolved from common ancestors, the distribution of a taxonomic group must have a history in which an ancestral species occupied a certain region and gave rise to various descendant species that may, or may not, have spread from there. Darwin used the geographic distributions of species as some of his most important evidence for evolution. For instance, he observed that species on islands are often much like species on the nearest mainland and usually are precisely those forms that are most able to disperse across salt water. Darwin inferred that the island species evolved from continental ancestors.

Alfred Russel Wallace, famed for having independently conceived the theory of evolution by natural selection, spent much of his life traveling and collecting specimens. He took a special interest in—and really founded—biogeography, the scientific study of the geographic distribution of organisms. He distinguished major biogeographic "realms" with fairly different sets of animals, such as the Palearctic (Europe and northern Asia), Nearctic (North America), Oriental (southern Asia), and Australian (including New Guinea and some islands to the west). "Wallace's line," east of Bali, marks a steep difference between a primarily Oriental bird fauna (such as woodpeckers) and one more characteristic of Australia and New Guinea (such as lorikeets and other parrots).

The distribution of a major taxon, such as a genus or family, often presents intriguing, sometimes hard questions: how did it get there, and (often) why is it only there and not elsewhere? Did hummingbirds always inhabit tropical America, or did they come from somewhere else and then diversify there? And why aren't they in Africa or Eurasia? Some groups are even more puzzling. If you have seen an Elegant Trogon (*Trogon elegans*) in Arizona or Mexico, and you then travel to central Africa or tropical Asia, you may see birds that you will immediately recognize as trogons (plate 39). And what about the giant flightless birds? Ostriches in Africa, rheas in South America, emus and cassowaries in Australia and New Guinea, and moas, only recently extinguished by humans, in New Zealand—how did they get to these remote, separated parts of the world?

Biologists have postulated four kinds of historical events that might explain the geographic distribution of a clade of species. The simplest explanation, particularly germane to groups that are restricted to a single region, is that the clade *originated* there and hasn't spread far. The owlet-nightjars, fairywrens, and birds-of-paradise are a few of the families that are found only in Australia and New Guinea and probably evolved there. Conversely, a key explanation is *dispersal*. The Cattle Egret (*Bubulcus ibis*) became established in South America in the 1930s, evidently having crossed the ocean from Africa. A large population of Barn Swallows has developed in Argentina since 1960, when six pairs were found breeding within the species' overwintering range.[2] A clade of species may have expanded by dispersal from the range of its common ancestor, with species proliferating by allopatric speciation in different colonized areas. That is, a widely distributed species may undergo *cleavage* ("*vicariance*") into two or more separate populations that evolve into different species. Some biologists have postulated a special case of vicariance, in which a clade may have once occupied a single land mass that was cleaved by continental drift into two or more lands, carrying the ancestors of today's species. Finally, *extinction* may have caused a clade's distribution today to differ from its past distribution.

How can we discover the historical cause of a group's distribution? For biological history, we look to the fossil record and to phylogeny, especially phylogenies based on DNA sequences, in which the rate of sequence divergence has been calibrated so that we can estimate when divergence occurred (recall chapter 2). Consider the Wrentit (*Chamaea fasciata*), a common bird in the chaparral vegetation of the Pacific coast of North America. Ornithologists recognized long ago that it is not closely related to any other North American birds. Using mitochondrial DNA of diverse passerines, researchers confirmed earlier suspicions that it is closely related to the *Sylvia* warblers of Eurasia, such as the Blackcap.[3] Clearly, the Wrentit has descended from an ancestor that expanded from Asia into North America, probably across the Bering land bridge between present-day Russia and Alaska—the route that humans also followed, much later.

Most of the studies I will describe use phylogenetic relationships to infer histories of bird distribution, but they aren't always the last word—especially because extinction can erase the historical record. It would be reasonable to suppose that hummingbirds originated in South America, since all hummingbirds are American and the great majority are South American. I will never forget my shock and astonishment one day in May 2004, when I opened a weekly issue of *Science* and saw this title: "Old World fossil record of modern-type hummingbirds." Gerald Mayr, a leader in bird paleontology, reported several unequivocal hummingbird skeletons from Oligocene (33.9–23.0 Ma)

FIGURE 11.1. A fossil stem-group hummingbird, *Eurotrochilus* sp., found in early Oligocene deposits in southeastern France. The skeleton and even feathers are unusually well preserved. (Courtesy of N. Tourment and A. Louchart, copyright Springer Nature, by permission.)

deposits in Germany.[4] Since 2004, other specimens have been found elsewhere in Europe, some showing early steps in the anatomical evolution of modern hummingbirds (figure 11.1).[5] The immediate common ancestor of living hummingbird species was probably in tropical America, but its ancestors may have been in Europe or elsewhere. This is by no means the only such example.[6] The fossil record in Europe also includes members of the Hoatzin family, of which only one species survives today, in South America. Turacos, restricted to Africa today, occurred in North America in the Eocene.[7]

Bearing this in mind, let us consider some puzzling disjunct distributions, such as trogons in tropical America, Africa, and southeastern Asia. One suggested explanation is continental drift: the idea that the ancestors of trogons inhabited the southern landmass Gondwana and that their distribution was cleaved when Gondwana separated into several land masses, where the descendants of these early trogons persist today. To judge this hypothesis, we have to know how and when Earth's land masses formed. It turns out that Gondwana has played some role in bird distributions but mostly in easing dispersal among land masses. The timing is important.

About 300 Ma, during the evolution of early reptiles, the Earth's land masses aggregated into a giant continent, Pangaea. During the efflorescence of the dinosaurs, in the Jurassic period (around 175 Ma), Pangaea broke up into northern and southern continents, Laurasia and Gondwana. Gondwana started its fragmentation when the Atlantic Ocean began to form between Africa and America. By 150 Ma, during the Cretaceous period, a land mass that is now Africa and South America started to separate from the rest of Gondwana, which included India, Madagascar, Australia, New Zealand, and warm, forested Antarctica. By the late Cretaceous, around 100 to 90 Ma, both Africa and a Madagascar + India land mass were fully separated from other land masses. Also in the late Cretaceous, a plate carrying New Zealand, New Caledonia, and other Melanesian land masses broke away from Australia, which itself started to separate from Antarctica about 80 Ma. Meanwhile, the South Atlantic Ocean had continued to form as South America moved westward from Africa. South America finally separated from Antarctica about 30 Ma. (This is when Antarctica became fully icebound.)

Both the fossil record of birds and the earliest divergence among living clades of birds (judged by the DNA "clock") agree that most Cretaceous birds became extinct during the great Cretaceous–Paleogene (K–Pg) extinction, 66 Ma, and that the few ancestors of most living birds started to diversify at or soon after that time (see chapter 3). Although Australia was still separating from Antarctica and South America was still connected to it, most of the separation of land masses by continental drift occurred before the K–Pg extinction (figure 11.2). This suggests that ornithologists should be careful about attributing the distribution of living birds to the fragmentation of Gondwana.

Take the trogons. An early phylogenetic study, based on the sequence of two genes, concluded that the common ancestor of the American, African, and Asian trogons lived in the Oligocene (40 to 23 Ma). In 2015, Richard Prum's team published a phylogenetic study of most of the living bird families, based on 259 genes (figure 3.6). They suggest an even more recent estimated divergence between American and African trogons, at about 20 Ma. Both of these dates are much too late for continental drift to explain trogons' distribution. What is more, all of the trogon fossils that have been found are in Europe. A likely explanation of trogons' distribution today is that they were widely distributed in Eurasia and North America, as well as to the south, and became extinct in the north as the climate cooled.[8]

One of the biggest surprises in the evolution of birds is our new understanding of ostriches and other giant flightless birds: the ratites. Ratites were presumed to have evolved from a giant ancestor that had lost the ability to fly. Because of this presumption and their distribution on southern land masses,

K–Pg Boundary 66 Ma

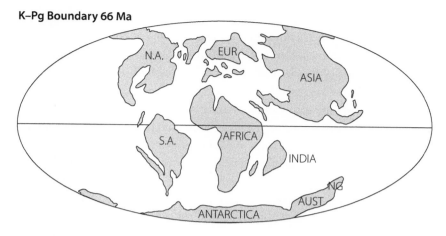

Middle Eocene 50.2 Ma

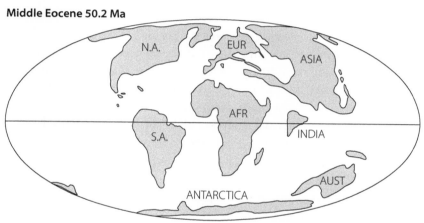

FIGURE 11.2. The distribution of major land masses at the Cretaceous (K)–Paleogene (Pg) boundary, 66 million years ago (Ma) (upper diagram) and in the Middle Eocene, 50 Ma (lower). Africa was already an isolated land mass 66 Ma, at the time of the K–Pg mass extinction and before the diversification of modern birds. By the Middle Eocene, Madagascar and India had separated, India was approaching Asia, and the continents were well separated, except that Australia was still close to Antarctica. (After maps in Paleomap Project, www.scotese.com.)

they became a textbook example of a distribution caused by continental drift.[9] That turns out to be wrong: their ancestors almost certainly flew to several land masses and became flightless giants afterward.

The oldest divide in the phylogenetic tree of living birds, based first on anatomy and later on DNA, is between the Neognathae (most birds) and the Palaeognathae.[10] The Palaeognathae include the Neotropical tinamous, vaguely chicken-like birds that can fly (plate 40), and the flightless ratites: ostriches in Africa, rheas in South America, kiwis in New Zealand, Emu (*Dromaius novaehollandiae*) in Australia, and cassowaries in Australia and New Guinea (which together are sometimes called Australo-Papua). Indigenous human populations extinguished two other families of ratites, the moas in New Zealand and the elephant birds in Madagascar. (These were eliminated so recently that you can see their eggs in some hotels in Madagascar.) All the ratites are or were gigantic, except the kiwis and some of the moas.

The ratites share some anatomical features, such as the absence of a keel on the breast bone, so they were traditionally classified as a monophyletic group that included all the ratites but not the tinamous. But in 2008, a large research team led by John Harshman concluded, from the sequences of twenty nuclear genes, that the tinamous are not the sister group of the ratites but instead fall within them. Subsequent studies confirmed this and used ancient DNA, extracted from bones, to find the phylogenetic position of the moas and the elephant birds as well.[11] It is surprising but almost certain that the tropical American tinamous are most closely related to New Zealand moas and that they are nested within the flightless ratites. Equally surprisingly, the New Zealand kiwis are related to the Madagascan elephant birds (figure 11.3).

This new phylogeny has major implications. First, because all birds descend from flying ancestors (chapter 3), and because tinamous fly, *either* (1) the ancestors of tinamous were flightless but tinamous re-evolved flight and associated anatomy, such as a breast keel, *or* (2) tinamous are descended from an unbroken line of flying ancestors, and flight has been independently lost in at least three lineages of ratites (as shown in figure 11.3). The authors of these studies agree that the second hypothesis—convergent evolution of flightlessness—is much more likely. One reason is that there are several other, unrelated, extinct groups of giant flightless birds: it seems to have been a common evolutionary path that was followed when, because of the K–Pg mass extinction, there were neither large dinosaurs (now extinct) nor large mammals (not yet evolved) filling the ecological niches that these birds evolved to occupy. Another reason has been recognized by evolutionary biologists for a long time: complex features that are lost during evolution almost never re-evolve in anything like the same form. (For example, none of the 3,000+ species of snakes have re-evolved legs, and no birds have re-evolved teeth; cf. figure 6.2.) The anatomy necessary for flight is certainly complex. Moreover, the ratite phylogeny itself suggests that some groups dispersed over water by flight and then became flightless. Kiwis, in New Zealand, and elephant birds, in

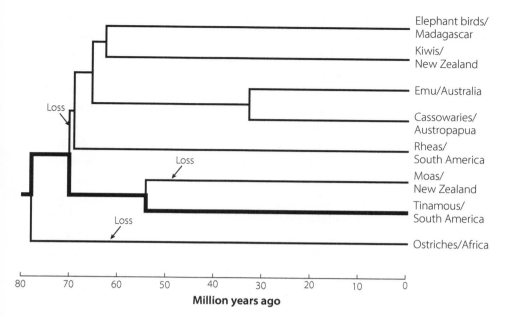

FIGURE 11.3. A molecular phylogeny of families of palaeognathous birds, the ratites and tina-mous. Two extinct groups, the moas (Dinornithidae) and elephant birds (Aepyornithidae), are included, based on DNA extracted from skeletal remains. The lineage leading to the tinamou family is shown in bold. We postulate that this lineage was able to fly because we know that all living birds are descended from flying ancestors. If the tinamou lineage retained flight through-out its history, flight was independently lost at least three times, indicated by arrows along the branches leading to ostriches, moas, and the ancestor of the rheas, elephant birds, kiwis, Emu, and cassowaries. Considering the great distances among Madagascar (with elephant birds), New Zealand (with kiwis), and Australia-New Guinea (with cassowaries and Emu), it is very likely that flight was independently lost in these groups as well. (After Yonezawa et al. 2017 and Cloutier et al. 2019.)

Madagascar, diverged from their common ancestor about 60 Ma, but both those land masses had separated from the rest of Gondwana more than 20 million years earlier. Descendants of a flightless ancestor couldn't have walked to both those regions. In fact, all the living groups of palaeognaths, in their separate lands, diverged after 70 Ma (at most), when most of the land masses were already separate.

Perhaps the best case that can be made for a role of continental drift in palaeognath history is suggested by fossil ostriches and other probable palae-ognaths that have been found in Europe and Asia. In this scenario,[12] the Pal-aeognathae, fully capable of flight, originated in the north and spread into

South America in the late Cretaceous (perhaps 80 to70 Ma). Because part of Gondwana was fairly intact at that time, palaeognaths could spread from South America into Antarctica, and then into Australia. As these ancestral paleognaths populated the new regions, many of them evolved large size and became flightless, filling similar, empty ecological niches. In this scenario, Gondwana explains these birds' distribution not by splitting and separating related birds into different regions, but by providing a route for dispersal, largely by flight.

Needless to say, ornithologists want to piece together the geographic history of as many bird species as possible, and, needless to say, there are uncertainties and debates in what is very much a detective work in progress. The most intense biogeographical research has concerned the Passeriformes—perching birds or songbirds—that make up about half of the world's bird species.[13] The first suggestion that they may have originated in the Australian region was by Charles Sibley (from whom I learned bird systematics as an undergraduate) and his colleague Jon Ahlquist. Long before DNA could be readily sequenced, they used a very crude method, called DNA–DNA hybridization, to assay the overall similarity between the DNA of different species.[14] On this basis, they proposed a phylogeny of many of the world's birds. Although the method and the ways the data were analyzed were pioneering efforts, we now know that many of their suggested relationships were wrong. However, they were the first to propose that a great variety of very different passerines in Australia—species that had been considered wrens, robins, flycatchers, thrushes, and more—form a single clade. Moreover, Sibley and Ahlquist suggested that this lineage, together with other Australian species such as lyrebirds, were the ancestral clade from which most of the world's passerines were derived and that Australia was the ancestral home of the Passeriformes. Crude as their analysis was by today's standards, it appears that Sibley and Ahlquist were largely right on this point.

Ornithologists agree on the basic phylogenetic framework of the Passeriformes (figure 11.4). Almost all the earliest branches in the phylogeny are among groups in the southern hemisphere, especially the Australian region, suggesting that the major clades of Passeriformes originated there. Anatomical and DNA data agree that the two living species of New Zealand wrens (Acanthisittidae) are the sister group to all other passerines, which form two great clades that diverged about 50 Ma, the suboscines (Tyranni) and the oscines (Passeri). The suboscines form two clades, the extremely diverse Neotropical Tyrannides (tyrant flycatchers, ovenbirds, antbirds, cotingas, and others) and the Eurylaimides, which include the broadbills (plate 41) and pittas in tropical

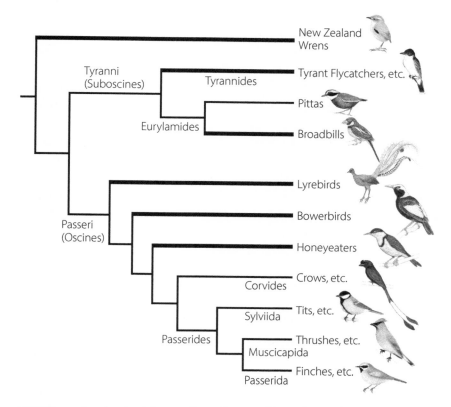

FIGURE 11.4. A molecular phylogeny of some lineages of Passeriformes, showing major groups mentioned in the text. The species figured are (from top down) New Zealand Rockwren, Eastern Kingbird, Banded Pitta, Black-and-red Broadbill, Superb Lyrebird, Regent Bowerbird, Banded Honeyeater, Greater Racket-tailed Drongo, Great Tit, Cedar Waxwing, Golden-winged Warbler. Thick branches are groups distributed mostly on Gondwanan land masses. (Based on Oliveros et al. 2019; art, Luci Betti-Nash.)

Africa and Asia—as well as a tropical American species, the Sapayoa (*Sapayoa aenigma*). These groups fit a Southern Hemisphere pattern that seems to implicate Gondwana. The oscines also seem to have originated in the Gondwanan region.[15] Their phylogeny includes several successive basal branches, such as the lyrebirds (see figure 2.3), bowerbirds (plate 42), fairywrens (see plate 2), and honeyeaters, before we get to two huge branches, the Corvides and the Passerides. All these basal families are found in Australia, New Guinea, or both. That screams "Australo-Papuan origin!" and it is, indeed, the conclusion reached by a team of leading systematists who analyzed the sequence of 4,060 genes in almost all (137) of the families of Passeriformes.[16] They conclude that the earliest lineages of passerines diverged in the Australo-Papuan region about 47 Ma, in the Eocene epoch, and radiated into the major clades during the Oligocene.

One of these clades, the Corvides (named from *Corvus*, the genus of crows and ravens)[17] includes about 700 species in more than thirty families. Among these, the shrikes (Laniidae) and crows (Corvidae) are familiar to birders in Europe and North America, while Europeans also know the European Golden Oriole (*Oriolus oriolus*) in the Oriolidae and Americans often encounter vireos (Vireonidae), which extend into South America. But the majority of families in this clade inhabit Australia and especially New Guinea.[18] These include birds-of-paradise, berrypeckers, whistlers, woodswallows, fantails, sitellas, quail-thrushes, butcherbirds, and other small families. Several research groups, using increasingly massive DNA data and more advanced statistical analyses, have concluded from the phylogeny that the Corvides originated in the Australo-Papuan region in the late Eocene.[19]

Some of the Australo-Papuan families, such as the fantails and whistlers, have extended their range into Wallacea (the islands between New Guinea and Borneo) and the Greater Sundas. The drongos are among the families that have dispersed further, into tropical Asia. Some lineages of the Corvides dispersed further still and gave rise to new families far from their Australo-Papuan origin: the beautiful bushshrikes (Malaconotidae) in Africa, the diverse Vangidae in Africa and Madagascar, and the shrikes (Laniidae) and vireos (Vireonidae) that I mentioned before. And the eponymous Corvidae are almost worldwide: ravens and crows in most of the world; nutcrackers and choughs in northern Eurasia; ground jays in the Asian steppes; treepies, blue magpies, and green magpies in tropical Asia; and a diverse array of exquisitely beautiful jays from Eurasia through tropical America. These are mostly very social, highly intelligent birds, and it is tempting to suppose that these features have been the key to their successful spread into diverse habitats around the world. In fact, there is evidence that bird lineages with larger brains have increased in diversity at a higher rate, probably by occupying and adapting to novel ecological conditions.[20]

The other great branch of the oscines, the Passerides, includes many of the songbirds most familiar to Europeans and North Americans, such as tits, swallows, thrushes, wrens, warblers, sparrows, and buntings. But the basal, earliest-branching families are mostly in southern lands, such as the Australo-Papuan satinbirds (Cnemophilidae) and the most peculiar rockfowl (Picathartidae) in western Africa. (Every "world birder" wants to see the rockfowl, and I have an unusually vivid memory of my experience of the Grey-necked Rockfowl [*Picathartes oreas*] in Cameroon [plate 43]. Groups of these lanky, bare-headed, crow-sized birds bound across the ground on long, strong legs and build mud nests on the sides of rock faces, which they visit every afternoon even when not actively nesting. Our group trekked through rain forest to a rocky slope with two gigantic egg-shaped boulders that leaned against each other, forming an open-ended cavern in which we ensconced ourselves as inconspicuously as possible to await the birds. Eight birds arrived; wary at first, they accepted our presence and were content to check their nests while we viewed and photographed. As we awaited them, I was attracting at least one hundred honeybees that found me more irresistible than my companions. After the rockfowl were accustomed to us, I slowly cast off my sweaty t-shirt, which helped to deflect the bees, although only after several stings. Fortunately, the birds paid no attention, and we enjoyed them for about a half hour.)

Most of the families of Passerides fall into three fairly distinct groups, the Sylviida, Muscicapida, and Passerida. The phylogenetic relationships among many of the families within these groups are uncertain because they originated within a short time in the late Oligocene and Miocene. The Sylviida are mostly in Africa and the Palearctic (Europe and northern Asia): they include tits, larks, bulbuls, babblers, swallows, leaf warblers, and sylviid warblers (and the American Wrentit that I mentioned earlier). The Muscicapida include Palearctic groups that have also spread into North America and even tropical America: waxwings and relatives, thrushes, nuthatches, creepers, dippers, Old World flycatchers, starlings, thrashers (in America only), and wrens (most diverse in Middle America).

The third major group, Passerida, has a most interesting distribution and history. The basal families are mostly in South Africa (sugarbirds) and tropical Asia (sunbirds, flowerpeckers, leafbirds). Then there is a great array of families that are often called "nine-primaried oscines" because their wings lack one of the ten primary feathers that are typical of almost all other birds. Among them, the more basal lineages are mostly in Africa (weavers, waxbills) or the Palearctic: accentors, Old World sparrows such as the House Sparrow, pipits and wagtails, and the finch family Fringillidae. A huge superfamily, Emberizoidea, is mostly American.

The finch family Fringillidae has taken some fascinating evolutionary turns (figure 11.5). This family includes Northern Hemisphere species such as Common Chaffinch (*Fringilla coelebs*), Hawfinch (*Coccothraustes coocothraustes*),

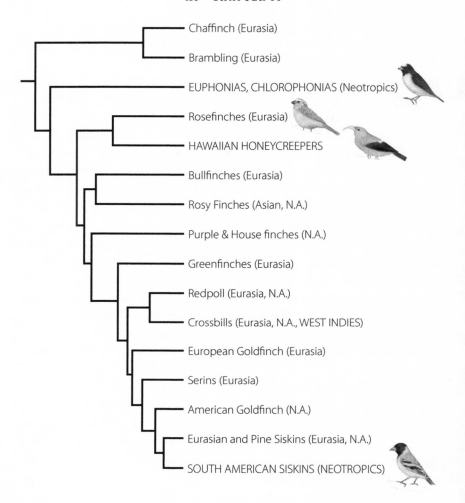

FIGURE 11.5. A phylogeny of some members of the finch family Fringillidae, showing the origin of the fruit-eating euphonias and chlorophonias, the Hawaiian honeycreepers, and the South American siskins. (After Zuccon et al. 2012; art, Luci Betti-Nash.)

crossbills, siskins, and Eurasian rosefinches (*Carpodacus*). But the family has expanded beyond northern latitudes at least three times. One event was quite recent: expansion of the siskins into tropical America, where there are about fourteen species that extend, especially in mountains, to the southern tip of South America.[21] Considerably earlier, another fringillid entered South America and gave rise to a highly modified group, the thirty-four recognized species of euphonias and chlorophonias—colorful fruit eaters that were thought to be tanagers until molecular phylogeny showed otherwise.[22] The third fringillid excursion takes us away from tropical America to the Hawaiian Islands, where the great adaptive radiation of Hawaiian honeycreepers (plate 44) evolved from a fringillid ancestor that was closely related to the rosefinches.[23] In many fringillid species, flocks undertake long-distance flights as they seek food, a lifestyle that predisposes them to colonize far-flung places.

The superfamily Emberizoidea, which arose about 20 Ma, includes the 44 species of Emberizidae, or Old World buntings, and about 790 species in the New World, mostly in five families: Cardinalidae (cardinals and relatives), Thraupidae (tanagers),[24] Passerellidae (New World sparrows), Icteridae (blackbirds and orioles), and Parulidae (New World warblers). A phylogenetic study,[25] based on sequences of six genes in almost all the species, showed that the ancestor of the Emberizoidea came from Eurasia to North America via the Bering route, founding a lineage that diversified into these five families (and some small ones as well) between 20 and 10 Ma. One of these, the Emberizidae, dispersed back into Eurasia and diversified there. Four lineages—the sparrows, blackbirds, warblers, and the stem lineage of cardinals and tanagers—diversified in North America and also dispersed into South America between about 11 and 7 Ma. During this period, these continents were widely separated by a strait in the region now bridged by the Isthmus of Panama, which slowly emerged and was fully formed about 2.5 Ma, at the end of the Pliocene epoch. The long separation probably prevented a wholesale movement of North American families into South America, so the lineages that did cross faced relatively little competition and could diversify freely. They certainly did: all five families became diverse, especially the Thraupidae, which underwent a spectacular adaptive radiation into about 375 species of fruit- and insect-eating tanagers, insect-gleaning conebills, nectar-feeding honeycreepers and flowerpiercers, and a great panoply of seedeaters and finches. The emberizoid families, especially the Thraupidae, account for most of the oscines in tropical America, complementing the great diversity of flycatchers, ovenbirds, and other suboscines.

When the Isthmus of Panama became complete at the end of the Pliocene, there transpired what has been called the Great American Biotic Interchange, which has been most fully described for mammals. North American groups such as the deer, cat, and squirrel families invaded South America, while North America received South American lineages, such as the opossum, an armadillo, and the now extinct ground sloths. Similarly, many bird lineages—both emberizoids, such as tanagers, and others, such as flycatchers and woodcreepers—crossed the bridge in both directions.[26] Some South American species are also found in Central America, having arrived there so recently that they have not yet become different species: you can see Blue-grey (*Thraupis episcopus*) and Bay-headed Tanagers (*Tangara gyrola*) and Red-legged Honeycreeper (*Cyanerpes cyaneus*) in both northern South America and Costa Rica. Conversely, the Rufous-collared Sparrow (*Zonotrichia capensis*), widely distributed in South and Central America, is a recently evolved member of the North American genus that includes the White-throated Sparrow.

———

The profuse proliferation of tanagers and other birds in tropical America illustrates the enormous variation in numbers of species around the world. My home state, New York, covers 141,297 square kilometers, with a rich variety of ecological biomes that range from ocean shore and a coastal plain with southern species (Chuck-will's-widow [*Antrostomus carolinensis*], Boat-tailed Grackle [*Quiscalus major*], Blue Grosbeak [*Passerina caerulea*]) to the Adirondack Mountains in the north, with boreal species such as Spruce Grouse (*Falcipennis canadensis*), Canada Jay (*Perisoreus canadensis*), and White-winged (Two-barred) Crossbill (*Loxia leucoptera*). Including many rare vagrant species (e.g., Kirtland's Warbler [*Setophaga kirtlandii*], Calliope Hummingbird [*Selasphorus calliope*], Ancient Murrelet [*Synthliboramphus antiquus*], Corn Crake [*Crex crex*]), 495 species have been recorded in the state, of which 248 are known to have nested.[27] A very active birder who travels throughout the state can hope to see about 335 species during a year; the record was set in 2012 by Anthony Collerton, whose huge effort was rewarded with 361 species.[28] In contrast, I have participated in birding tours that recorded 512 species in northern Peru, 520 in central Kenya, 602 in Uganda, and an astonishing 692 in Bolivia—each of these within three to four weeks! Of course, these countries are bigger than New York, and in Peru and Bolivia the Andes Mountains provide a great many habitats for different species—but three of these regions lack a sea coast, with its great array of shore-living and pelagic species. A species count from a year in the Rocky Mountains would not begin to approach the numbers in Peru or Bolivia. How can we explain such variation?

Mountainous regions generally support more species than extensive flat lands, probably for three reasons. One is obvious: the great array of vegetation types and climatic conditions at different altitudes provides many potential ecological niches. A species' ecological niche is the range of combinations of different environmental conditions, such as temperature, and different resources, such as food and nest sites, that enable its population to persist. Species with excessively similar requirements are unlikely to coexist indefinitely because of competition (see chapter 10), so more species can be "packed" into an area that offers more niche options.

The other explanations for the high diversity in mountainous regions are more historical. For one, mountains offer more opportunity for populations of a species to become isolated in different valleys or on different peaks and to become different species. In other words, the rate of speciation is higher than in flatter areas where birds encounter fewer barriers and can maintain gene flow over long distances. As well as enhancing opportunities for speciation, mountains may also have reduced rates of extinction: as climates fluctuate, a bird species may continue to prosper by shifting to a different elevation that provides the temperature and rainfall regime to which it is already adapted.[29]

The most impressive and endlessly debated pattern of variation in biological diversity is the "latitudinal diversity gradient." I have been intrigued by this problem ever since the 1960s, when I started to teach tropical ecology courses in Costa Rica. In almost every major group of animals and plants, the number of species is greatest in tropical latitudes, steadily diminishing in temperate and polar regions. We might suppose that the cause is obvious: life is easier where it is warm and wet than where it freezes. But you wouldn't say that if you were a Snowy Owl or a walrus: it depends on your evolutionary history of adaptation. If most of the Earth had had Toronto's or Stockholm's climate for most of Earth's history, we might expect most species to be cold adapted.

Another seemingly obvious explanation might be that tropical regions receive more solar energy and so have a higher growth of plant biomass: higher productivity that can support more insects, more birds, more species. But how come the greater energy is partitioned among more species? Why isn't it garnered by fewer but more abundant species of plants, insects, or birds? (One of the most remarkable features of birding in tropical rainforests is how astonishingly uncommon, or downright rare, many of the species are. A knowledgeable birding guide often will take you to one of only a few accessible sites where the species can be dependably seen. In Uganda, a group of us descended 250 meters of altitude, walking for two hours, to see a family of Grauer's Broadbills [*Pseudocalyptomena graueri*] at what we were told is the only known site, anywhere, for reliably seeing this species. This is an extreme example, but many other tropical species are nearly this hard to see.)

There are three basic models of why one region (A) has more species of a clade, such as birds, than another region (B). We envision that over time, in each area, the number of species increases from one. The growth in number is exponential at first, but it eventually must level off as finite resources (food, space) limit the number that can coexist (figure 11.6). In model 1, both regions have attained their limit, but region A can support more species than B (possibly because of higher productivity). In model 2, the number of species is still growing in both regions, but the number grows faster in A than in B. This could be because birds in region A have a higher speciation rate (the probability that a species splits into two species during a given time interval) or a lower extinction rate. In model 3, the number of species is also still growing, but region A started to accumulate species before region B. Maybe the earliest species in region A arrived there from somewhere else before the earliest species in B; maybe region A became suitable for these birds before B did. Model 1 ascribes the regional difference in diversity to some ongoing, constant property of the environment. In models 2 and 3, history plays a more significant role.

Based on many studies, especially using phylogenetic information, it seems that all these models help to explain the latitudinal diversity gradient in birds.

There is a lot of evidence that the number of species in most ecosystems is at or near a limit—that ecological niches are mostly full.[30] (Most established introduced species do not contradict this because they tend to use novel human-created environments—and some reduce populations of indigenous species.) Some evidence lies in the shape of time-calibrated phylogenies: they commonly show a great burst of divergence of different lineages for a short period early in their history, with fewer and fewer new lineages being added over time. An example is a subfamily (Fluvicolinae) of tyrant flycatchers, which ancestrally lived mostly in Neotropical forests and shifted into more open habitats as these expanded in the Miocene.[31] They underwent an explosive diversification in size, shape, behavior, and habitat use, as their names suggest: shrike-tyrants, water tyrants, chat-tyrants, and the remarkable ground tyrants, long-legged, pipit-like inhabitants of barren ground high in the Andes (figure 11.7). The fluvicoline phylogeny shows a great initial burst of speciation and morphological divergence, followed by a decline in the accumulation of new species—except for a late, Pleistocene burst of speciation in the ground tyrants. We also know that the number of species in a clade

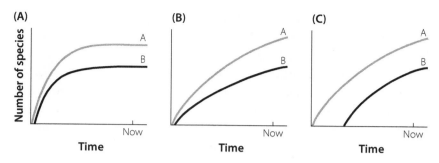

(A)

Number of species

A

B

Now

Time

(B)

A

B

Now

Time

(C)

A

B

Now

Time

FIGURE 11.6. Three models of how one region, A, can come to have more species than another, B. (A) Both regions have a stable (equilibrium) number of species; more can coexist in region A than B because of some environmental difference, such as the variety of food resources. (B) Species diversity is still increasing in both regions, but faster in region A, either because of a higher rate of speciation or a lower rate of extinction. (C) Species diversity is increasing at the same rate in both regions, but region A has more species because the accumulation of species started earlier. (After Mittelbach et al. 2007.)

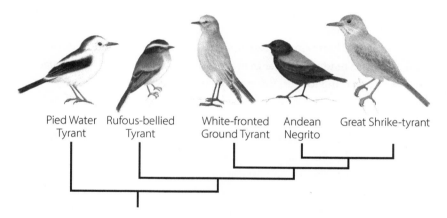

Pied Water Tyrant   Rufous-bellied Tyrant   White-fronted Ground Tyrant   Andean Negrito   Great Shrike-tyrant

FIGURE 11.7. Some genera that illustrate adaptive radiation in the flycatcher subfamily Fluvicolinae. Left to right: a water tyrant (*Fluvicola*) that feeds mostly on the ground in wet areas; an insectivorous chat-tyrant (*Ochthoeca*) that lives in shrubby habitat; a ground tyrant (*Muscisaxicola*) that runs about fairly bare open ground; a negrito (*Lessonia*), on the ground near water; and a shrike-tyrant (*Agriornis*), a large, open-country flycatcher that captures large insects and small vertebrates. (Phylogeny after Fjeldså et al. 2018; art, Luci Betti-Nash.)

often has increased precisely when it invades a new region where ecological opportunities await—as when the tanager-cardinal lineage entered tropical America from North America, or when a rosefinch colonized the Hawaiian Islands and gave rise to species diversely adapted to feeding on insects, fruits, and nectar.[32]

If both temperate-zone and tropical environments have reached their species limits, why would the limit be higher in the tropics? One possibility is that birds can be more diverse because their food sources—plants, insects, and vertebrate prey—are more diverse. For example, Neotropical forests harbor more "sit-and-wait" predators, such as puffbirds, jacamars, and motmots, that wait for large insects, lizards, or other good-sized prey that are much more common there than in high-latitude forests. And there is certainly a correlation between bird species diversity and plant productivity, which is greatest in warm, wet tropical environments and might support more species (as in model 1). Exactly how it would do that is uncertain, and it is hard to separate the effect of productivity from other processes.[33]

What about rates of speciation and extinction (model 2)? There is little evidence that rates of speciation are higher in tropical than temperate species. If anything, phylogenies seem to show that a greater fraction of high-latitude species evolved by recent speciation, which seems faster there than in the tropics.[34] However, Michael Harvey, Brian Smith, and their colleagues have found that the amount of genetic difference among populations of a species is greater within tropical than temperate-zone species of birds.[35] They propose that tropical species tend to be older than higher-latitude species: their populations have been diverging longer. Eventually, such populations may evolve reproductive isolation and become different biological species. However, Smith and colleagues found that the rate of population divergence is three times greater than the rate of speciation. This implies that most of the populations that become genetically differentiated do not become species. Many divergent populations—nascent or ephemeral species—become extinct, because of environmental change or because they come into contact, interbreed, and fuse, not having evolved strong enough reproductive isolation.[36]

Might a lower rate of extinction explain high tropical diversity (model 2b)? The genetic evidence that tropical species are older suggests this possibility. Time-calibrated phylogenies also offer some evidence (although estimating extinction rates from phylogenies is very tricky). For example, a "long branch"—a lineage that separated long ago but has only a few, very closely related, recently formed species—is one that probably spawned many species at various times in its past that became extinct. This is one of several possible signals that enable researchers to compare a time-calibrated phylogeny with computer simulations that assume various rates of speciation and extinction, and then to see which model fits the data best. Paola

Pulido-Santacruz and Jason Weir took this approach in a study of several large clades (such as Passeriformes) in both temperate and tropical regions.[37] They concluded that extinction was the key factor: in all cases, the tropical branches apparently had a history of lower extinction rates. This difference in extinction is consistent with other information. The "basal" branches of many bird families—descendants of ancient ancestors—are more often tropical species, whereas the temperate-zone members often are recently evolved, closely related species.[38]

One possible cause of a tropical–temperate difference in extinction rates might be a less stable climate at higher latitudes. Climates change every 10,000 to 100,000 years, partly because of "Milankovich cycles": changes in Earth's orbit due to gravitational effects of other planets, especially Jupiter and Venus. These climate changes are likely to cause shifts in species' ranges, as some populations of a species become extinct and new sites are colonized. These processes break down the genetic differences that have developed among different populations of a species, setting back progress toward speciation and undoubtedly causing some species to become extinct.[39] The climatic changes caused by Milankovich oscillations are more pronounced at higher latitudes.

In a longer historical frame, tropical or subtropical conditions prevailed throughout most of the world before the Oligocene, and groups that are restricted today to the tropics, such as the mousebird and turaco families that I mentioned earlier, occurred in Europe and North America.[40] These groups did not adapt to the colder, more seasonal climate that started to expand over much of Earth in the Oligocene; they were set in their ways, so to speak, and became extinct except at lower latitudes, where their environment persists today.

This history supports the idea that the tropics may have more species simply because species have been accumulating longer in tropical than in temperate or polar environments (model 3). One clue to this possibility is that tropical regions have not only more species: they also have more genera and families of birds. In other words, living in tropical climates has deep, old phylogenetic roots. Almost all major clades of plants and animals, including birds, originated in a tropical climate because that climate enveloped most of Earth until less than 35 Ma. Adaptation to colder, more seasonal climates may not be easy. (Neither is adaptation to a hotter climate, as I will explain in chapter 12.) And so birds, like other organisms, show "phylogenetic niche conservatism," tending to remain associated with their ancestors' environment and

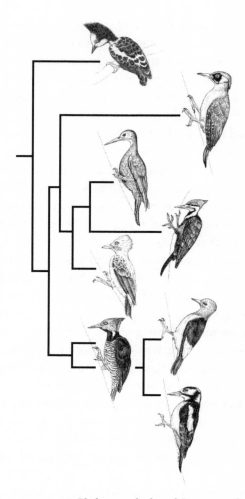

FIGURE 11.8. Phylogeny of a few of the genera of woodpeckers, illustrating repeated origin of high-latitude species or genera from tropical ancestors. The tropical species, from top to bottom at left, are Heart-spotted (*Hemicircus canente*), Great Slaty (*Mulleripicus pulverulentus*), Cream-colored (*Celeus flavus*), and Guayaquil (*Campephilus gayaquilensis*) Woodpeckers. The high-latitude species, from top to bottom at right, are Eurasian Green (*Picus viridis*), Pileated (*Dryocopus pileatus*), Red-headed (*Melanerpes erythrocephalus*), and Great Spotted (*Dendrocopos major*) Woodpeckers. (Phylogeny based on Shakya et al. 2017; art, Stephen Nash.)

way of life: rather few groups have adapted to the novel rigors of cool, seasonal climate, and most of these rather recently. This conclusion appears to be a major explanation of tropical diversity, not only of birds, but of trees, amphibians, reptiles, and mammals as well.[41]

Of course, parts of Earth have been cool or arid ever since birds first evolved, and some lineages have adapted to these environments throughout bird history. Fossils show that penguins have inhabited the southern continents since the Cretaceous and have evolved as the Antarctic region became more frigid. Some species in groups that had evolved in warm environments adapted to polar climates, as did various owls, woodpeckers, and many other groups (figure 11.8). Likewise, the Australian outback supports species of parrots, honeyeaters, and other birds that adapted as much of the continent became more arid. It is hard to think of an environment or a resource that an air-breathing animal might use but that birds have left vacant.

And so it is that whether we travel in person or vicariously through books, films, or Internet, we can enjoy and marvel at special birds in every region. We may want to know how they came to be and why they act and look the way they do; we may simply enjoy their colors, songs, behaviors; we may look each year for familiar species and greet them as old friends; we may set ourselves challenges such as listing; we may feel a kind of spiritual union with nature's plenitude: these are the pleasures and the enrichment of our lives that we may find in the evolved, and evolving, living world.

12

Evolution and Extinction

THE FUTURE OF BIRDS

Since 1970, bird populations in North America have declined by 29%, a net loss of 3 billion birds (figure 12.1). At least 40% of the world's species of birds are declining, and 1,469—1 in every 8 species—are threatened with complete extinction.[1] Birds are not alone in peril. Rhinoceroses, pangolins, and tigers are only some of the best-known endangered mammals; many amphibians, reptiles, and pollinating insects are in decline; freshwater species face pollution, dams, and drought; marine fisheries threaten both target and nontarget species; corals are dying, and with them an extraordinary diversity of fishes and other animals that live in reefs. In the 541 million years since animals first diversified, there have been five so-called mass extinctions, when great numbers of species perished. Many biologists are convinced that we are now in the first stages of the sixth mass extinction, one uniquely caused by the actions of a single species.[2] Understanding today's decline and extinction of bird species, and what conservation measures might be effective, is largely a study of birds' ecology, together with the economics, politics, and sociology of human affairs. In this chapter, I ask what understanding may be gained from the perspective of evolutionary biology, which has tried to understand both extinction and survival of species in the past and the ability of species to adjust or adapt to changes in their environment.

I encourage readers to look at *The State of the World's Birds*,[3] a report of BirdLife International (BLI), for a comprehensive view of the causes of bird extinction, conservation actions past and present, Important Bird Areas, and the status of every species, classified by the International Union for the Conservation of Nature (IUCN) as Critically Endangered (222 species), Endangered (461), Vulnerable (786), Near-Threatened (1,017), or Least Concern

(A)
(B)

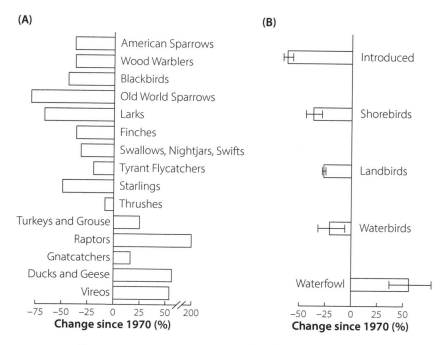

FIGURE 12.1. Changes in North American bird abundance since 1970, categorized by (A) taxonomic families and (B) introduced species and ecological (management) groups. Bars extending to the left of the zero lines indicate declines; those to the right, increases (mostly in waterfowl, raptors, and [inexplicably] vireos). (After Rosenberg et al. 2019.)

(8,417).[4] (Table 12.1 provides a sample of threatened species for several regions.) I use "threatened" to refer to the first three categories.

Among the causes of anthropogenic (human-caused) extinction, the BLI report lists agriculture as having the biggest impact (on 74% of the 1,469 threatened species), followed by logging (50%), invasive species (39%), and hunting (35%).[5] Climate change and associated severe weather threaten a huge number: about 33% of species. Agriculture and logging are the chief causes of habitat loss or degradation; agriculture includes not only clearing for grains and vegetable crops but also massive deforestation for cattle ranching (to feed the excessively beef-rich diet of the most affluent countries) and for oil palm plantations, which are replacing forest en masse in tropical Asia. Agriculture also kills birds via pesticide use, now including especially neonicotinoids. The invasive species that have threatened or extinguished the most birds are predators, especially cats and rats, but goats have affected birds by devastating

TABLE 12.1. Examples of Critically Endangered and Endangered Species in Several Regions and Countries*

	Status	
	Critically endangered	Endangered
North America (U.S. and Canada)	California Condor, Ou (Hawaiian Islands)	Gunnison Sage Grouse, Black Rail, Whooping Crane, Marbled Murrelet, Saltmarsh Sparrow
Mexico	Short-crested Coquette, Guadalupe Storm Petrel, Cozumel Thrasher	Horned Guan, Thick-billed Parrot, Azure-rumped Tanager
Dominican Republic	Ridgway's Hawk	White-fronted Quail-Dove, Black-capped Petrel, Hispaniolan Crossbill
Europe	Sociable Lapwing, Yellow-breasted Bunting	White-headed Duck, Egyptian Vulture, Northern Bald Ibis, Saker Falcon
China	Baer's Pochard, Chinese Crested Tern, Blue-browed Laughingthrush	Hainan Peacock-Pheasant, White-eared Night Heron, Jankowski's Bunting
Indonesia	Sumatran Ground Cuckoo, Helmeted Hornbill, Straw-headed Bulbul, Bali Myna	Maleo, White-winged Duck, Java Sparrow
Japan	Okinawa Woodpecker	Okinawa Rail, Izu Tit, Red-crowned Crane, Blakiston's Eagle-Owl
India	Great Indian Bustard, Indian Vulture, Slender-billed Vulture	Forest Owlet, Greater Adjutant, Saker Falcon
Kenya	White-headed Vulture, Hooded Vulture, White-backed Vulture	Black-fronted Francolin, Grey Crowned Crane, Grey Parrot
South Africa	White-winged Flufftail, White-headed Vulture	African Penguin, Black Harrier, Botha's Lark
Brazil	Brazilian Merganser, Glaucous Macaw, Araripe Manakin	Wattled Curassow, Sun Parakeet, Buff-throated Purpletuft
Peru	Junín Grebe, Waved Albatross, Royal Cinclodes	Titicaca Grebe, Royal Sunangel, Golden-backed Mountain Tanager
Australia	Plains-wanderer, Swift Parrot, Orange-bellied Parrot, Regent Honeyeater	Golden-shouldered Parrot, Night Parrot, Noisy Scrub-bird, Forty-spotted Pardalote
New Zealand	Black Stilt, Kakapo, Black Robin	Southern Brown Kiwi, Yellow-eyed Penguin, Takahe, New Zealand Rockwren

Source: Information from BirdLife International, State of the World's Birds (2018).

vegetation, and introduced parasites and pathogens have also been important. (Human-introduced bird malaria has extinguished some Hawaiian honey-creepers, and an introduced parasitic fly threatens some Darwin's finches in the Galápagos Islands.[6]) Staggering numbers of birds are directly killed or captured. The toll in the eastern Mediterranean, where birds are eaten, ranges from 2.3 million birds per year in Cyprus to 5.5 million in Egypt and 5.6 million in Italy. In tropical Asia, millions of birds, of hundreds of species, are captured for the cage-bird trade, and several species, such as Java Finch (*Lonchura javanica*) and Straw-headed Bulbul (*Pycnonotus zeylandicus*), are Critically Endangered as a result. In the United States, billions of birds fall prey to domestic cats, and many millions more are killed by colliding with building glass, automobiles, communication towers, and power lines.[7]

We should also be very concerned about an ecological change that will have a huge impact on many species of birds and on many other aspects of terrestrial ecosystems. Throughout the world, but documented especially in North America and Europe, insect populations are declining dramatically in what has been called an "insect apocalypse."[8] One careful study in Germany documented a *75% decline* (!) in the biomass of flying insects during twenty-seven years. Formerly common British species of butterflies and moths have declined rapidly. There is strong reason to think that both climate change and modern agricultural practices, especially the use of pesticides, are partly to blame, and while the causes may not be fully understood, "we know enough to act now."[9] As in other kinds of organisms, insect species vary greatly in abundance, so there is fear that a great many of the less common species (possibly 40%) may be at risk of extinction. The implications for birds are of course profound, since so many species depend on insects—including many seed eaters that feed insects to nestlings. Recent population declines of aerial insectivores such as swallows may be related to insect decline.[10]

Population declines and extinction vary among environments and among bird taxa. On a global basis, more species are threatened or extinguished by the loss of forest, especially tropical forest, than other habitats simply because these habitats are so species-rich. In terms of percentage of species, grassland birds in the United States and elsewhere have suffered worst. Wetland species have increased in the United States, but in many parts of the world the situation is very different. Migratory species of birds have declined more than non-migratory species, partly because of habitat loss in their wintering areas or en route.[11] For instance, migratory shorebirds are severely affected, in China and elsewhere, by the loss of intertidal sites for resting and refueling (plate 45).

Overall, the bird species in greatest danger are probably those on small islands, where many are endangered or already extinct, chiefly because of habitat loss. The bird taxa that have suffered the greatest threat (leading to extinction in some cases) include parrots and cockatoos, cranes, albatrosses and some other seabirds, and vultures.

In chapter 4, I described several examples in which birds have rapidly adapted, by natural selection, to changes in their environment. What is the prospect that bird species can adapt to the great dangers that BLI describes? Can evolution rescue species from extinction? It depends on the nature of the threat. I think it unlikely that birds can substantially adapt to hunting and trapping for food or the cage-bird trade. There is little doubt that many bird species are more wary of humans where they have been hunted for many generations than where there has been little history of hunting. Darwin and other visitors to the Galápagos Islands have been impressed by how tame most of the birds are, while birders who have been to New Guinea know how hard it is to see species that have been hunted for thousands of years. But those hunters are still very successful: humans are wilier enemies than any others that birds have ever faced, and I doubt natural selection can do much about it.

What about habitat loss? This is more complicated. First, bear in mind that natural selection, the mechanism of adaptation, is based on differences in fitness among individual organisms, not on any benefit to the population. If populations survive, it is because individuals do. In most species, the population is fairly stable from year to year, not growing. That is because males or pairs divide suitable terrain into defended territories, and most of the terrain is fully occupied most of the time. Consequently, most young birds fail to find vacant ground, fail to breed, and probably die without reproducing. It ought to be advantageous to be flexible, and use a different habitat, but Scarlet Tanagers, nightingales, bowerbirds, and other forest-dwelling species don't settle down and breed in hedgerows or suburban yards.[12] So if half the forest is cut down, more birds experience insufficiency of habitat—but this isn't a novel environmental change for the population, which in every generation has included great numbers of young birds faced with exactly this insufficiency. As more and more forest disappears, the bird population dwindles, but it is just as nicely adapted to its forest habitat as it was before, and there may be no *novel* natural selection for adapting to any other habitat, even as the population reaches the brink of extinction.

Obviously, bird populations have adapted to different habitats in the past, but adaptation to a very different habitat is difficult if it requires many different adjustments in how to feed, escape predators, nest, raise young, and contend with novel competing species It is doubtless for this reason that related species of birds often occupy quite similar habitats, illustrating what has been called phylogenetic niche conservatism.[13] Almost all larks live on the ground in open country, most wood warblers in forest, almost all antbirds in forest understory. Australian honeyeaters originated in wet, subtropical forests, and only a few branches of the honeyeater phylogeny became adapted to drier habitats, over millions of years, as Australia became more arid.[14] Quite a few species thrive in human-altered landscapes, but most of them occur "naturally" in physically similar habitats, often second-growth vegetation with a mixture of trees and brush. In tropical America, you will see Tropical Kingbirds (*Tyrannus melancholicus*) and Blue-black Grassquits (*Volatinia jacarina*) in both second growth and farmland, but antshrikes, antpittas, and manakins only in fairly mature forest.

Evolution might be most likely to rescue bird species that are threatened by some invasive species and by climate change. In chapter 4, I described how House Finch populations rapidly evolved resistance to a bacterial pathogen, so we can have reasonable hope that many bird species in the future will adapt to novel pathogens or parasites. This has undoubtedly happened many times in the history of birds—although we will never know how many bird species failed to adapt to these threats. Adaptation to highly proficient predators like cats may be a different matter and may vary greatly among kinds of birds, their modes of escape, and their habitats.

Climate change, caused by human release of greenhouse gases, is the most severe, global, long-range threat to the biosphere, as well as to human welfare. The world's average temperature has already risen by 1°C, may increase by another 1.5°C by 2050 at the current rate, and may well increase by a devastating 3°C by the end of this century unless greenhouse gas emission is severely reduced.[15] The increased heat generates violent weather events and severe droughts that result in calamitous fires of the kind seen recently in California, Brazil, and Australia. Large areas are projected to attain summer temperatures beyond the tolerance limits of humans and many other species, and severe drought has already caused the collapse of arid-land bird communities in Australia and the Mojave Desert of the United States.[16] The rise in ocean level endangers nesting birds on some atolls, as well as certain coastal birds, such as the Saltmarsh Sparrow (*Ammodramus caudacutus*) on the Atlantic coast of North America.

Climate change may impact bird populations in several ways. High temperatures may simply exceed the tolerance of birds at one or another life stage.

Bird species have declined in the Mojave Desert because with recently increased temperatures, they cannot get enough water for evaporative cooling.[17] An important impact is increased asynchrony between the phenology (seasonal schedule) of birds' life cycles and other species, especially insects and other food. This has been best described for migratory birds, which must time their spring arrival and egg laying to the seasonal abundance of food. Spring arrival of migratory passerine species in eastern North America has become earlier, as has leafing out ("green-up") of the foliage on which insects depend—but the birds' arrival has not kept pace and has increasingly lagged behind green-up.[18] In Europe, those populations of Pied Flycatchers in which nesting increasingly lags behind maximal food availability have declined by 90%.[19] On their tropical wintering grounds, migratory species lack information about the weather in their nesting area, and so individuals cannot adjust their migration schedule to the progress of spring in the north.

Not all populations and species suffer equally from climate change. A study of sixty-two European species found that the rate of population growth was lowest in populations closest to the warm (south) end of the species' range and increased in populations further north.[20] As we will shortly see, some species are likely to survive by shifting their geographic range, following the geographic shift in the climate regime to which they are already adapted. In mountainous regions, species' distributions will shift upward, but species already at the top will have no refuge and are at great risk of extinction. (Some mountaintop populations have already been extirpated in the Peruvian Andes.[21]) Tropical species may be at greater risk than high-latitude species.[22] Many are already living near their thermal limit. Also, the temperature gradient is shallower in the tropics than at high latitudes: you have to go much farther north or south if you are in Brazil than if you are in the central United States to find an equivalent difference in average temperature. Tropical species are less likely to shift their range far enough to stay in their optimal temperature regime. Bird populations are declining in at least some intact Amazonian forest, far from human disturbance—possibly because of climate change.[23]

Populations become extinct due to climate change (or any other change) if population growth is negative: if the death rate exceeds the birth rate. The population can persist either by shifting its distribution to a more favorable location where growth is positive or by adjusting in situ. Two kinds of adjustment might occur: by individual phenotypic plasticity or by adaptive evolutionary change. Phenotypic plasticity is expressed by alterations of an individual's physiology, behavior, or morphology (chapter 5). The alteration may be due to the direct effect of environment (such as the effects of child-hood nutrition) or the individual's behavior (compare the differences be-tween trained athletes and many of the rest of us). Phenotypic plasticity has limits and will suffice for population persistence only if the environment isn't too extreme. Plasticity is usually an adaptation that has evolved by natu-ral selection in a varying environment. For example, tropical lizards and insects have narrower thermal tolerances than temperate-zone species, which experience a far greater range of temperatures.[24]

Adaptive evolutionary change, the increase of favorable alleles and geno-types by natural selection (chapter 4), is called "evolutionary rescue" if it saves a population from extinction.[25] Adaptation must happen fast enough to change a population's growth from negative to positive before the population becomes so rare that it is doomed. I described in chapter 4 that many charac-teristics can evolve fairly rapidly because they are genetically variable, such as bill size in Darwin's finches. This is "standing variation," an accumulation of mutations that happened in the past and to which new mutations may be added, at a low rate, each generation. (At the same time, some old mutations are lost by genetic drift.) Large populations usually have more standing varia-tion, and they gain more new mutations each generation. For these reasons, and simply because a large population usually takes longer to dwindle to ex-tinction than a small population, evolution is more likely to rescue larger populations than small ones from climate-induced extinction. But there are many potential obstacles to evolutionary rescue. Survival and reproduction may depend on changing several different characteristics together, some of which may have little standing variation. Also, many potentially advantageous genes have harmful "side effects "(recall color polymorphisms, chapter 5) and will not contribute to adaptation unless their advantage outweighs their disadvantage.

Most of the research on possible adjustment to climate change has focused on body size and on breeding schedule. In chapter 1, I mentioned Bergmann's rule: the tendency for bird (and mammal) populations in colder regions to have larger body size than populations in warmer regions. Larger size reduces the loss of body heat because it reduces the ratio of surface area to body mass, so natural selection favors larger size in cooler climates—but smaller size in warmer climates. Body size has declined in a number of bird species, but in most cases the mechanism—evolution or phenotypic plasticity—is not known.[26] One of the few exceptions is a study of eight passerine species in southeastern Australia, most of which vary gradually (clinally) from smaller size in the (warmer) north to larger in the (cooler) south. Measurements of live birds and old museum specimens show that the entire gradient has shifted in the last one hundred years, so that southern populations now have the body size that was typical of northern populations before 1990. The authors provide evidence that the shift is not due to individual plasticity but to genetic change.[27]

Many species are known to have changed their breeding schedule. One of the most comprehensive studies uses data from the Grinnell Resampling Project.[28] Joseph Grinnell, the first director of the Museum of Vertebrate Zoology at the University of California in Berkeley, is renowned for his extraordinarily detailed, organized field notes, including those that he and his team took on a series of transects in California from 1911 to 1940. He was very aware that these data would be important for future studies. Ecologists have been revisiting Grinnell's sites and recording similar data from 2003 to 2010. These data show that among 224 species of birds, breeding phenology is 5 to 12 days (average = 8.6) earlier than in Grinnell's time. The shift may correspond to the availability of insect food for nestlings, and it also reduces thermal stress: nesting birds experience average temperatures at least 1°C cooler than if they had not shifted.[29] Among the many other species that are known to have shifted breeding time, most probably have done so by phenotypic plasticity; there are very few well-documented cases of genetic, evolutionary change.[30] One of the few exceptions is the Pied Flycatcher (figure 12.2). In 2002, Barbara Helm and colleagues repeated exactly the same experiment, with the same population of flycatchers, that had been done in 1981: they collected nestlings from the wild on the same dates, raised them under the same captive conditions, and measured differences in phenology.[31] Compared with 1981, the 2002 birds had earlier moult at the end of winter, earlier *Zugunruhe* (migratory restlessness), earlier activation of testes, and earlier egg laying (by nine days, compared with an eleven-day advance in the wild population). Because experimental birds experienced the same environment in 1981 and 2002, the differences are almost certainly based on genes, hence on evolution by natural selection.

Some specific genes are known to affect the timing of migration and breeding activity, such as a gene called *Clock* that is important in daily physiological cycles and other biological rhythms. *Clock* was first described in fruit flies, and it operates in birds, mammals, and probably all animals.[32] Variants of this gene differ in the number of repeats of a base pair triplet that encodes a certain amino acid, so some alleles are longer than others. The same variation is found in several other genes, such as one called *ADCYAP1*. In many species of birds,

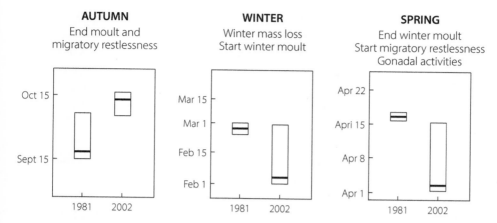

FIGURE 12.2. Changes in the seasonal cycle of a population of Pied Flycatchers (*Ficedula hypoleuca*), based on birds reared under the same conditions in 1981 and 2002. In each panel, the horizontal bar in a rectangle shows the median date in 1981 (left rectangle) and in 2002 (right rectangle). The rectangles indicate variation among individual birds. For each season, the data show averages of (left to right) the end of post-juvenile molt and of autumn *Zugunruhe*; winter mass loss and start of molt; and end of molt, start of spring *Zugunruhe*, and gonad activation. Data for males are shown; females showed similar differences. The autumn index shows that the end of molt and migratory restlessness in male birds was later in 2002 than in 1981. The winter and spring indices are earlier in 2002 than in 1981. (After Helm et al. 2019.)

variation in gene length is correlated with the timing of migration and breeding schedule.[33] In the Blackcap, which has evolved changes in migratory activity and direction (recall chapter 4), birds with longer *ADCYAP1* alleles have higher migratory activity; female Eurasian Blue Tits with shorter *Clock* genes breed earlier; Blackpoll Warblers with longer *Clock* genes arrive later in the spring at their breeding area, and those with shorter *ADCYAP1* genes arrive earlier at their wintering grounds.[34] In some species, then, there is variation in critical genes that might enable evolutionary changes in their seasonal activities as climate change alters the optimal timing.

On the whole, however, there is only a little evidence that birds are adapting to current climate change by evolutionary change. Instead, their most common response is probably "niche tracking": shifting their distribution to match the geographic shift of the climate to which they are adapted. Fossils of plants and animals from the Pleistocene glacial and interglacial episodes show that this has been a very common response; in fact, paleontologists often use plant species as climate indicators because their "climatic niche" seldom evolved.[35]

Niche tracking is a reflection of the phylogenetic niche conservatism that will be familiar to many birders. North American high-latitude birds, such as juncos, the Hairy Woodpecker (*Leuconotopicus villosus*), and the genus of the Hermit Thrush (*Catharus*), can be seen in Costa Rica, but only at high, cold elevations. Black Grouse and Rock Ptarmigan (*Lagopus muta*) are found in northern Europe—and in the Alps, doubtless persisting there since the glaciers retreated. In the Northern Hemisphere, many southern species have been extending their range northward. At least twelve southern species, such as Glossy Ibis (*Plegadis falcinellus*), American Oystercatcher (*Haematopus palliatus*), Red-bellied Woodpecker (*Melanerpes carolinus*), and Northern Mockingbird (*Mimus polyglottos*), were rare in the vicinity of New York City before the 1950s but now breed regularly or commonly.[36] In mountainous areas around the world, species are shifting to higher elevations, with tropical species perhaps tracking their niche more closely than temperate-zone species.[37] Using data from the Grinnell Resampling Project, Morgan Tingley and colleagues found that forty-eight of fifty-three species in the Sierra Nevada closely tracked their climatic niche.[38]

A shift in a species' distribution, in contrast to an expansion, involves both the colonization of new sites at the advancing edge of the range (upslope in mountains, northern in the Northern Hemisphere) and extinction of populations at the trailing edge (low altitude, or southern). John Wiens searched publications that reported the status of trailing-edge populations of animals and plants, mostly low-altitude populations in mountainous terrain.[39] He found that extinction of such populations has occurred in 23 of 61 temperate-zone species (37%) and 108 of 210 tropical species (51.4%). The data included several studies of birds, such as an analysis of birders' data in the New York State Breeding Bird Atlas, carried out in 1980–1985 and again in 2000–2005. Among 43 species with a southern range border in New York, the average range border retreated northward by 11.5 kilometers.[40]

This is worrisome, as it implies that populations are not adapting and that survival of many species may depend on dispersal into new areas. The ability of species to do that will vary. As I mentioned earlier, tropical species in broad expanses of lowland forest would have to shift a very long way to encounter cooler locations. A major concern is that humans have destroyed the "corridors" of suitable habitat that many birds need if they are to spread. This is a greater problem for an earthbound amphibian or millipede than for many birds, but broad urbanized or agricultural landscapes are severe barriers for many bird species, especially those that do not migrate, and more especially for countless species in tropical forests. For some of these, even a moderately broad river is a major barrier, much less fields of rice, corn, or sugar cane that extend for kilometers. Many bird species in northeastern Brazil, for example, are limited to small, widely scattered forest reserves—islands in an ocean of sugar cane—and are almost surely destined for extinction. Thanks to an outstanding Brazilian guide, I saw one of these species, the Alagoas Foliage-gleaner (*Philydor novaesi*, plate 46), in 2008. The IUCN declared it extinct in 2018.

It is hard to be optimistic about the likelihood that evolutionary adaptation will rescue many species from anthropogenic climate change. Ignacio Quintero and John Wiens estimated the average rate of niche evolution among closely related species of many vertebrates, including five bird clades, by comparing differences between their climatic niches relative to their age (based on a molecular clock). They concluded that the projected rate of climate change from now until 2100 is 10,000 to 100,000 times faster than rates of evolution of climatic niches. The rate of evolution might be higher for short periods than the average rate over the several million years that typically separate related species, so a similar study was done of variation among populations of individual species that live in climatically different regions.[41] Here, again, the projected rate of climate change was more than 200,000 times faster than the rates of evolution of the climatic niche. That still doesn't mean that evolution will necessarily be too slow to save species: the fossil record shows many examples of very low rates of evolution averaged over long time spans but punctuated by very brief periods of rapid change. Maybe some species will experience a rapid burst of evolutionary adaptation to climate change. The critical questions are, which characteristics need to evolve, if the population or species is to survive? And how much genetic variation in these features is available for natural selection to craft a new, improved bird that can tolerate a stressful new environment? That is a research project that has only just begun.

What can you and I do? First, despite the somewhat discouraging picture I have painted, do not despair. Climate change is a serious long-term threat, but at least for the near future it doesn't endanger two-thirds of bird species, according to BirdLife International. The immediate ongoing dangers, such as habitat loss, invasive species, and direct exploitation by hunting and trapping, threaten about 12% of species. These threats can be mitigated, at least sometimes, and have been successfully reduced in many cases: conservation efforts often do work. Such efforts have moved about twenty-five species, such as Lear's Macaw (*Anodorhynchus leari*) and the Crested Ibis (*Nipponia nippon*), from Critically Endangered to less dire status (plates 47, 48). Many other species have benefited from the creation of preserves, habitat restoration, and extirpation of invasive predators, such as the rats that threatened the South Georgia Pipit (*Anthus antarcticus*), the southernmost passerine species. Captive breeding efforts have rescued the California Condor (*Gymnogyps californianus*), the Whooping Crane, and other species. The IUCN is developing a Green List of recovering species, to complement its Red List of threatened species.[42]

These efforts, achieved mostly by diverse conservation organizations that sometimes act together, require action in the political, legal, economic, social, and educational realms. Actions range from bringing lawsuits against governments, to fundraising for nature preserves, to working with landowners or villagers in order to build management programs that benefit both birds and people. Some of us have skills that we can contribute to these efforts; others may help by joining and contributing to organizations. Most of us, in one way or another, can educate by showing other people, children or adults, the richness and beauty of species that share our environment, whether rural or urban. Bird-related organizations are prominent in conservation because people can see birds, then become interested in them, and then realize the dangers to birds and to whole ecosystems. To really see birds—or lizards, insects, plants, the bounteous life around us—is the first step toward caring.

> What would the world be, once bereft
> Of wet and of wildness? Let them be left,
> O let them be left, wildness and wet;
> Long live the weeds and the wilderness yet.[43]

NOTES

Chapter 1. In the Light of Evolution: Birds and Evolutionary Science

1. Species names are capitalized in birding literature (although not in scientific literature). Thus, a Common Tern (*Sterna hirundo*) is a particular species, whereas any of many species might be the "common tern" in a certain region. I cite a species' scientific name when it is first mentioned because different books and other sources use different English names for some species.

2. Smith 1990.

3. von Holt et al. 2018.

4. Thanks to Oscar Wilde.

5. Dobzhansky 1973.

6. Morphology refers to form—the physical features of an organism, including anatomical structures such as bones, as well as feathers, pigmentation, and so on.

7. The citations spanned from 2018 back to the 1870s. I don't know how complete the coverage is of literature before the mid-twentieth century.

8. Models in evolutionary biology are conceptual representations of biological systems that are intended to help us understand how those systems work. Many models are verbal or mathematical schemes of what might be the important elements in biological processes. For example, mathematical models that include rates of birth, death, immigration, and emigration might be used to predict changes in the size of a population. Verbal and mathematical models of natural selection describe ways in which characteristics or genes may evolve; they help us understand what factors affect evolution but seldom make precise, long-term predictions, in part because they deliberately exclude factors that are thought be less important, and they do not include all the particulars that are germane to a specific species.

9. The cited studies of House Sparrow are Johnston and Selander 1964, 1971; Bokony et al. 2006; Teplitsky et al. 2008; Brommer et al. 2014.

10. Buchanan and Cockburn 2013 and special issue of *Emu*; Cockburn et al. 2017.

11. Dennett 1995.

12. "Chance" is sometimes used in everyday speech to mean "without purpose or design," but it is strictly used in science to mean "random," or only probabilistically predictable. We cannot predict the identity of a card drawn from a thoroughly shuffled deck, but we can state that the chance of its being a diamond is one-fourth. Chance and randomness are philosophically very complex topics.

13. Birkhead et al. 2014.

14. Darwin 1839. "The Voyage of the Beagle" is the popular name, bestowed in 1905, for the book that Darwin wrote under a much longer title; see the bibliography.

15. The diverse breeds of domestic pigeons that fascinated Darwin are now studied with modern methods of genomics and developmental biology, for example, Domyan and Shapiro 2017.

16. The interpretation of Bumpus's (1899) data on House Sparrows has been the subject of some disagreement. See, for example, Buttemer 1992.

17. Lack 1947a.

18. Mayr 1942.

19. Sibley and Ahlquist 1990.

20. Nice 1937.

21. Readers who want more comprehensive information on birds and ornithology can find many fine books. Two of the most comprehensive are *Ornithology*, 4th ed., by Frank B. Gill and Richard O. Prum (MacMillan, 2019) and *The Cornell Lab of Ornithology Handbook of Bird Biology*, 3rd ed., edited by Irby J. Lovette and John W. Fitzpatrick (John Wiley and Sons, 2016).

Chapter 2. Parrots, Falcons, and Songbirds: The Bird Tree of Life

1. The meaning of "species" is a subject of unending discussion among biologists; see chapter 10. A species, in most of this book, is a group of populations that can freely interbreed among themselves but which interbreed and exchange genes little, or not all, with other such sets of populations.

2. James et al. 2003; Johansson et al. 2013.

3. Gill 2007.

4. Ellegren 2013.

5. Dawkins 1976.

6. Kachroo et al. 2015.

7. Quinn et al. 1991.

8. Ottenburghs et al. 2016.

9. Hackett et al. 2008; Jarvis et al. 2014; Burleigh et al. 2015; Prum et al. 2015.

10. Burns et al. 2014.

11. I don't describe one process because it gets too complicated. This is when a gene tree may differ from the corresponding species tree owing to "incomplete lineage sorting." Even the term is awkward! Among many examples is a phylogenetic study of Indigo Bunting (*Passerina cyanea*) and its relatives by Carling and Brumfield (2008).

12. Galla and Johnson 2014.

13. Lovette and Bermingham 1999.

14. Lovette et al. 2010.

15. To complicate matters, two Neotropical barbets have been designated a fifth family of their own, again because of the principle of monophyly.

16. The fifty-eighth supplement to the checklist, published in 2017, is by Chesser et al. The phylogenetic study of the sandpiper family is by Gibson and Baker 2012.

Chapter 3. After *Archaeopteryx*: Highlights of Bird History

1. Alan Feduccia and a few other ornithologists disagree with the consensus. See Feduccia, *Romancing Birds and Dinosaurs* (2020).

2. Field et al. 2020.

3. Ksepka et al. 2013.

4. Synonymous mutations and some noncoding DNA sequences that are thought to be neutral (not affecting fitness) are thought to account for most DNA sequence differences among species. The rate at which neutral mutations are fixed in a species depends on the rate at which they arise and fluctuate in frequency by genetic drift (figure 1.3). Both of these processes occur at about the same rate over long periods of time, so the replacement of old DNA base pairs by new ones should be roughly constant: clocklike. In contrast, DNA changes that affect fitness (survival and reproduction) are not expected to evolve at a constant rate. For instance, they might evolve faster when environments change rapidly.

5. Fleischer et al. 1998; Lerner et al. 2011.

6. Smith and Ksepka 2015.

7. Despite the theory, molecular clocks vary among distantly related groups. Many studies have attempted to account for the variation; for example, the clock tends to run faster for species that have short generation time. These rates give us a fair idea of when phylogenetic splits happened, but such estimates have a fair margin of error.

8. Mayr (2017) provides a comprehensive treatment of the fossil record of the origin and later history of birds.

9. The several groups of quadrupedal dinosaurs, such as *Triceratops*, evolved from bipedal ancestors.

10. The keel is lacking in Palaeognathae, the group that includes ostriches; see the description of bird phylogeny and classification, below.

11. Ostrom 1975.

12. Padian and Chiappe 1998.

13. Wiemann et al. 2018.

14. Prum and Brush 2002.

15. See Seebacher 2003 and Clarke 2013 on the debate about endothermy (metabolic production of body heat) in dinosaurs.

16. Xu et al. 2014; Brusatte et al. 2015.

17. Brusatte et al. 2015.

18. Wang and Lloyd 2016.

19. Field et al. 2018b. Bhullar et al. (2015) showed that altering key genes in chick embryos produces phenotypes like ancestral forms rather than beaked birds.

20. Louchart and Viriot 2011. Nonfunctional "fossil" genes, called pseudogenes, are found in almost all species' genomes, including that of humans.

21. Ksepka et al. 2017; Field et al. 2018a.

22. Alfaro et al. 2018.

23. Jarvis et al. 2014; Claramunt and Cracraft 2015; Prum et al. 2015.

24. Field et al. 2018a.

25. "Aequor" is a Latin word for sea, and "ornithos" is Greek for bird. Classical taxonomists seldom mixed these languages.

26. Ericson et al. 2006; Hackett et al. 2008.

27. A third suborder, Acanthisitti, has only two living species; see figure 2.3.

28. These names are capitalized when they refer explicitly to the taxa but usually not when used as nouns in sentences, rather like the use of "Mammalia" and "mammal."

29. Schluter 2000.

30. Michel et al. 2018. The gut microbiome of this finch resembles that of predators.

31. Cooney et al. 2017a.

32. Cooney et al. 2016.

33. Smith 2012.

34. Simpson 1964; Losos 2017. Stephen Jay Gould (1989) also developed this theme in his book *Wonderful Life*.

35. The three species of skimmers feed by flying with the uniquely elongated, compressed lower mandible in the water, closing rapidly on intercepted fish.

36. That makes two safe but unscientific predictions.

Chapter 4. Finches and Blackcaps:
How Bird Populations Change and Adapt

1. Lovejoy 1936. *The Great Chain of Being* was Lovejoy's history of the *scala naturae*, the "scale of nature" that theologians and philosophers, for centuries, held to be God's design.

2. Gregory 2009.

3. Cook 2003; Cook et al. 2012.

4. This is one of several formulations of fitness; some other, more complex measures are appropriate, depending on factors such as whether the species has sexual or asexual reproduction, and is stable or growing, or whether the environment is stable or fluctuating.

5. Baião et al. 2007.

6. More properly, the regression of the offspring's value on that of their parents.

7. Boag 1983.

8. Smith and Dhondt 1980.

9. Knief et al. 2012.

10. The average body size of survivors was about one standard deviation greater than non-survivors. The Browns and their collaborators have carried out many other studies in their long-term monitoring of this species. For example, colonies often nest along roads, and the Browns have found that roadkills have been declining, relative to the population size. At the same time, average wing length has declined in the population, whereas it is greater in road-killed samples. The Browns speculate that birds with shorter wings can more quickly lift off the road surface (where they often sit) and so escape oncoming vehicles (Brown and Brown 1998).

11. Much of this work is described in the Grants' several books, including *How and Why Species Multiply* (2008) and *40 years of Evolution* (2014).

12. Bonneaud et al. 2011.

13. Warner 1968.

14. Ongoing as I edit this text.

15. Berthold 1995.

16. A typewriter was a machine with keys bearing raised letters that struck an inked ribbon, impressing the ink letters onto a sheet of paper. A correction paper was a film with white material; an error in ink could be covered by white, and so hidden, by striking the same key onto the correction sheet, thus allowing the correct letter to be entered in its place.

17. Plummer et al. 2015.

18. Bezzel and Jetz 1995; Pulido et al. 2001; Pulido and Bethold 2010.

19. This German word is commonly used by biologists throughout the world who study bird migration. It means, literally, "movement (or migration) unrest."

20. This is the procedure by which humans have bred new varieties of domestic plants and animals.

21. Møller 2001; Marrot et al. 2018.

22. Lai et al. 2019.

23. Books by Mary Jane West-Eberhard (2003) and Sonia Sultan (2015) review phenotypic plasticity.

24. Yeh and Price 2004; Atwell et al. 2014.

25. The behavioral differences could not be studied in these birds because this species will not breed in captivity. However, the behavior is strongly correlated with testosterone level among birds within a population and is almost certainly directly affected by it.

26. A good introduction to cultural evolution, as it applies to humans, is Richerson and Boyd, *Not by Genes Alone* (2005). Whiten (2019) reviews cultural evolution in nonhuman animals.

27. Fisher and Hinde 1951; Aplin et al. 2013.

28. Aplin et al. 2015.

29. Turbott 1970; Pierce 1979.

30. Schondube and Martínez del Rio 2003.

31. A clear overview of this subject, comprehensible to readers who, like me, have forgotten all the physics they might have learned, is provided by B. W. Tobalske in *The Cornell Lab of Ornithology Handbook of Bird Biology*, edited by Lovette and Fitzpatrick 2016.

32. Rensch 1947, 1959.

33. For shorebirds, see Minias et al. 2015; for a broader sample of European birds, see Vagasi et al. 2016.

Chapter 5. The Ruff and the Cuckoo: Variation within Species

1. Witmer 1996; Hudon et al. 2013, 2017; Toews et al. 2017.

2. Smeds et al. 2016.

3. Cooke et al. 1988.

4. Balakrishnan and Edwards 2009.

5. A SNP is correlated only with other genes that are close by (linked) on the chromosome because more distant genes become separated by recombination during the formation of egg and sperm cells. The two corresponding chromosomes that an individual inherited from its parents swap parts, and the more widely spaced two genes are on the chromosome, the more likely they are to be swapped.

6. Briskie and Mackintosh 2004.

7. Forrest and Naveen 2000.

8. Hatching failure in New Zealand birds is described by Briskie and Mackintosh 2004. The 2008 population size of Kakapo is cited in Wikipedia, from https://kakaporecovery.createsend1.com.

9. Kruuk et al. 2002.

10. Much of the research on these crossbills has been done by Craig Benkman, at the University of Wyoming, and his associates (e.g., Benkman 2003; Snowberg and Benkman 2007).

11. This form of natural selection is often called sexually antagonistic selection or intralocus sexual conflict, and it can be very effective in maintaining genetic variation within a population. Incidentally, sexually antagonistic selection can strongly constrain or prevent evolution of a feature that is advantageous to one sex because the alleles that enhance the feature cannot increase in frequency if they are highly disadvantageous to the other sex

12. Tarka et al. 2014.

13. Overviews of color polymorphism in birds include Galeotti et al. 2003; Roulin 2004; Roulin and Ducrest 2013. Broader reviews of the genetics and adaptive functions of vertebrate pigmentation include Ducrest et al. 2008; Hubbard et al. 2010.

14. Pryke and Griffith 2009; Kokko et al. 2014.

15. Lowther 1961; Tuttle 2003; Tuttle et al. 2016.

16. Horton et al. 2014.

17. van Rhijn 1991.

18. Küpper et al. 2016.

19. Ducrest et al. 2008.

20. Passarotto et al. 2018.

21. Emaresi et al. 2013, 2014.

22. Sumasgutner et al. 2016.

23. Krüger et al. 2001; Kappers et al. 2020.

24. Chakarov et al. 2008.

25. Some overviews are *Cuckoos, Cowbirds and Other Cheats*, by N. B. Davies (2000), and reviews by Rothstein 1990; Davies 2011; Feeney et al. 2014; Soler 2014.

26. Brood parasitism has evolved independently in three lineages of the cuckoo family (Payne and Sorensen 2005).

27. Some Australian birds recognize and can discriminate against nestling bronze cuckoos. Most host species don't, possibly because birds form a mental image of their offspring (they are "imprinted") and might reject their own young if they were to first see and become imprinted on a parasite nestling.

28. Gibbs et al. (2000) studied the cuckoo; Spottiswoode et al. (2011) the honeyguide.

29. Rothstein 1990.

30. Brooke and Davies 1988. The Dunnock (*Prunella modularis*), which does not reject cuckoo eggs, was an exception.

31. Caves et al. 2015; Spottiswoode and Stevens 2010, 2012.

32. Soler 2014.

33. Lovászi and Moskát 2004.

Chapter 6. Hoatzin and Hummingbirds: How Adaptations Evolve

1. Prum et al. 2015.

2. Kornegay et al. 1994; Mackie 2002; Godoy-Vitorino et al. 2008.

3. Baldwin et al. 2014.

4. Place 1992; Place and Stiles 1992; Downs et al. 2002.

5. Witmer and Van Soest 1998; Witmer and Martínez del Rio 2001.

6. Tibetan populations' adaptation to high altitude is one of the best-documented cases of recent evolutionary adaptation in humans. They have a distinct allele of the *EPAS1* gene, which upregulates production of red blood cells and has other advantageous effects. See Yi et al. 2010; Beall 2014.

7. Natarajan et al. 2018.

8. McCracken et al. 2009; Zhu et al. 2018; Lim et al. 2019.

9. Prum 1999; Prum and Brush 2002.

10. On the evolution of novel morphological features, see Wagner 2015.

11. See de Bakker et al. 2013; Towers 2017. The homology (correspondence) between the fingers of birds, dinosaurs, and other tetrapods has been the subject of some controversy. I have accepted the very interesting hypothesis by Wagner and Gautier 1999; see also Vargas and Wagner 2009.

12. Recall from chapter 3 (and figure 3.6) that one of the earliest branches in bird phylogeny is between Galloanseres (including chicken) and the Neoaves (most birds, including doves and finches). That digit I starts to develop in chickens suggests that Galloanseres may retain more of the ancestral developmental process than Neoaves.

13. Brumfield et al. 2007.

14. Abzhanov et al. 2004; Mallarino et al. 2011; Silva et al. 2017.

15. Manegold and Töpfer 2012. The innermost tail feathers of *Hemicircus* are much like those of typical woodpeckers; the outer pairs are like those of piculets.

16. Temeles et al. 2009; Maglianesi et al. 2014.

17. Abrahamczyk et al. 2014.

18. Pauw et al. 2008; Muchhala and Thomson 2009.

19. The broadest review of the evolution of specialization, although a little dated, is Futuyma and Moreno 1988.

20. MacArthur 1958.

21. Martin et al. 2017; Robinson and Terborgh 1995.

22. Minot and Perrins 1986; Gustaffson 1987; Török and Tóth 1999; Martin and Martin 2001; Wiens et al. 2014.

23. This and related issues are the subjects of two excellent books, *The Ecology of Adaptive Radiation*, by Dolph Schluter (2000), and *Evolution's Wedge: Competition and the Origins of Diversity*, by David Pfennig and Karin Pfennig (2012).

24. The terms *sympatric* and *allopatric* will be important in the chapter on speciation.

25. Fjeldså 1983.

26. Mayr 1960; Wyles et al. 1983.

27. You can find amazing videos of this bird by entering "New Caledonian Crow" in a search engine such as Google.

28. Lefebvre et al. 1957; Sol et al. 2005; Ducatez et al. 2020.

29. Werner and Sherry 1987.

Chapter 7. Owls and Albatrosses: Life Cycle Events and Variations

1. AnAge: The Animal Ageing and Longevity Database, genomics.senescence.info.

2. Fay et al. 2016.

3. A slightly dated, but still excellent book on the evolution of birds' life histories and mating systems is *Evolutionary Ecology of Birds*, by P. M. Bennett and I. P. F. Owens (2002).

4. Lack 1947b. Perrins (1964) reported the Swift experiment.

5. Dijkstra et al. 1990; Stearns 1992.

6. Gustafson and Sutherland 1988.

7. Prince et al. 1992.

8. Weimerskirch 1992.

9. Robinson et al. 2010.

10. Death by predation is commonly greater for very young than for older birds but is thought to be a relatively constant risk during much of the adults' life span.

11. George Williams (1966) theorized that because of the physiological costs of reproduction, genes that increase reproductive effort early in life have the side effect of reducing reproductive effort later in life. Experiments with fruit flies have confirmed this idea.

12. Reed et al. 2008.

13. Boonekamp et al. 2014.

14. Moreau 1944; Skutch 1949; Robinson et al. 2010.

15. Young 1994.

16. Remeš et al. 2012.

17. Jetz et al. 2008; Robinson et al. 2010; Muñoz et al. 2018.

18. Martin 2015.

19. West-Eberhard 1983.

20. Lyon et al. 1994.

21. Götmark and Ahlström 1997.

22. My treatment of this topic draws on an enjoyable book, *More Than Kin and Less Than Kind*, by Douglas Mock (2004), an authority on the subject.

23. Bruce 1999.

24. Mock 2004, p. 71ff.

25. Bortolotti et al. 1991.

26. Anderson 1990.

27. Robillard et al. 2016.

28. Cockburn 2006.

29. Lyon et al. 1987.

30. I wish I didn't have to write this, because it sounds like a biological justification for the worst kind of sexist behaviors in human societies. But it doesn't justify any human behavior; it is the stark algebra of natural selection, just like the algebra of siblicide and overproduction of offspring. As I wrote earlier, don't look to nature for moral or ethical guidance.

31. Kokko and Jennions 2008.

32. Sexual selection is a major topic in the next chapter.

33. Székely 1996; Székely and Cuthill 1999; Székely et al. 1999; Székely et al. 2004.

34. Liker et al. 2013.

35. Olson et al. 2009; Remeš et al. 2015.

36. Griffith et al. 2002; Lyon and Eadie 2008. Other species are mentioned in chapter 5.

37. Sorenson 1991.

38. Gowaty and Bridges 1991; Sorenson 1991; Lyon 1993.

39. Brown and Brown 1991.

40. Lyon 2003. Lyon also discovered that coots stop laying when the clutch size reaches a certain number and that (remarkably) they can count eggs, apparently using visual cues. So they can determine if an extra egg has appeared in the nest and then search and destroy.

41. Feeney et al. 2014.

42. Harris et al. 2014.

Chapter 8. Auklets' Crests and Peacocks' Trains: Sexual Selection in Birds

1. Berglund et al. 1996.

2. West-Eberhard 1983; Lyon and Montgomerie 2012.

3. Velmala et al. 2015.

4. del Hoyo et al. 1992, vol. 1, p. 529.

5. Emlen 2014.

6. Verkuil et al. 2014; Gibson et al. 1991.

7. Studd and Robertson 1985.

8. Møller 1988.

9. Andersson 1982.

10. Petrie et al. 1991, 1992; Petrie and Halliday 1994; Loyau et al. 2005; Dakin and Montgomerie 2013.

11. Latin *extra* means "outside of."

12. Mulder et al. 1994.

13. Sheldon and Ellegren 1999; Michl et al. 2002; Whittingham and Dunn 2016.

14. The Zebra Finch study is by Forstmeier et al. (2011), the Blue Tit study by Kempenaers et al. (1997).

15. Clark and Dudley 2009. But Jennions et al. (2001) analyzed published studies of forty species of birds, spiders, insects, and fish, and concluded that, overall, more highly ornamented males tend to have higher survival, not lower. They suggested that vigorous males are the ones that can develop more elaborate traits and can "pay the cost."

16. Ryan 2018.

17. Gomez and Théry 2007.

18. The experiment was by Jones and Hunter (1998); the phylogeny is from Smith and Clarke (2015) and earlier studies.

19. Burley and Symanski 1998; Moehring and Boughman 2019. It has been suggested that experimentally attached ornaments of this kind might simply be perceived as increasing the males' size.

20. Zahavi 1974.

21. Rowe and Houle 1996.

22. See chapter 4.

23. Hamilton and Zuk 1982; Hill 1991; Hill and Farmer 2005.

24. Ligon and Zwartjes 1995.

25. Zuk et al. 1995; Parker 2003; Parker and Ligon 2007.

26. Hasselquist et al. 1996.

27. Somewhat different, complementary versions were formulated in the 1980s by Russell Lande and Mark Kirkpatrick. I've enjoyed stimulating interactions with both, and Mark and I have coauthored the most recent edition (2017) of the textbook *Evolution*.

28. Remember that both sexes inherit genes that govern the features of both sexes, even though they may be active in one sex

29. Zuk et al. 2006.

30. There is experimental evidence for this cost in Pied Flycatchers (Slagsvold et al. 1988).

31. Kuijper et al. 2012.

32. Prokop et al. 2012.

33. The ornithologist Richard Prum enthusiastically advocates the runaway hypothesis in *The Evolution of Beauty: How Darwin's Forgotten Theory of Mate Choice Shapes the Animal World—and Us* (2017). His book has been well received in the general press but criticized by some researchers in this area (e.g., reviews by Borgia and Bell 2018; Patricelli et al. 2018).

34. Darwin 1871, part 2, p. 277.

35. Lande 1980.

36. Potti and Canal 2011.

37. Clutton-Brock 2007; Kraaijeveld et al. 2007; Tobias et al. 2012; Hare and Simmons 2019.

38. Roulin et al. 2001. This is part of a more complicated story because male coloration is also correlated with offspring success, as more reddish males provide more food.

39. Heinsohn et al. (2005) studied the Eclectus Parrot; Ekstrom et al. (2007) studied the Greater Vasa Parrot.

40. Jones and Hunter 1999.

41. Tobias et al. 2011. The species were Yellow-breasted and Peruvian Warbling Antbirds (*Hypocnemis subflava* and *H. peruviana*).

42. The phylogenetic analyses are by McGuire et al. (2014) for hummingbirds and by Burns et al. (2014) and Shultz and Burns (2017) for tanagers. The godwit study is by Schroeder et al. (2009).

43. Friedman et al. (2009) studied orioles, and Simpson et al. (2015) wood warblers.

44. Holland and Rice 1998.

45. Brennan et al. 2007; Prum 2017. The larger theme is the subject of *Sexual Conflict*, by G. Arnqvist and L. Rowe (2005).

46. Coker et al. 2002.

Chapter 9. Anis, Swallows, and Bea-eaters:
The Social Life of Birds

1. Koenig et al. 1995.

2. Charles Wilson, a CEO of General Motors who became a US Secretary of Defense, is sometimes misquoted as saying that it is (Wikipedia).

3. Some books on this chapter's topic include those by Stacey and Koenig 1990; Ligon 1999; Koenig and Dickinson 2004.

4. Cockburn et al. 2017.

5. Brown 1986, 1988. Beauchamp (1988) reviewed such studies on many species and concluded that food intake rates commonly are higher the larger the foraging group—especially for broadly dispersed resources, such as seeds.

6. Creswell 1994a; Quinn and Cresswell 2006.

7. Carrascal and Moreno 1992; Sridhar et al. 2009; Griesser 2013.

8. Goodale et al. 2020.

9. Martínez et al. 2018.

10. Flower et al. 2014.

11. Cresswell 1994b.

12. Krams et al. 2008; Krama et al. 2012.

13. DuVal 2007.

14. Krakauer 2005.

15. Hatchwell 2009; Green et al. 2016.

16. Cockburn 2006.

17. Jetz and Rubinstein 2011.

18. Feeney et al. 2013.

19. Hatchwell et al. 2013.

20. Reyer 1990.

21. Komdeur 2003; Brouwer et al. 2012; Spurgin et al. 2014; Groenewoud et al. 2018.

22. Recall (chapter 4) that new mutations and other rare alleles are lost by genetic drift faster in small than large populations, so the level of genetic variation depends on population size—and can actually be used to estimate the population size (making a few assumptions).

23. Bonier et al. 2007; Navara 2013.

24. Emlen 1990, p. 514.

25. Riehl 2013.

26. Vehrencamp and Quinn 2004; Riehl 2011; Riehl and Strong 2018, 2019.

Chapter 10. Bird Species: What Are They and How Do They Form?

1. Titles and URLs are in the bibliography under Clements Checklist 2019, Gill and Donsker 2019, and Dickinson et al. 2014, respectively.

2. Barrowclough et al. 2016.

3. Brewer 2018.

4. Chesser et al. 2010.

5. de Queiroz 2007; Winker 2010.

6. Two or more species that look almost identical are sometimes called "sibling species" or "cryptic species."

7. Stresemann 1919.

8. Mayr 1942, p. 120.

9. See IOC checklist (Gill and Donsker 2019).

10. Price 2008; Uy et al. 2018.

11. Baker and Baker 1990.

12. Ratcliffe and Grant 1983.

13. Recall from chapter 3 that passerines are divided into suboscines (such as tyrant flycatchers) and oscines (such as thrushes and finches), which have a more complex syrinx. Suboscines' songs are innate (genetically based), not learned.

14. Sætre and Sæther (2010) describe such a case in the European Pied Flycatcher.

15. Parkes 1951; Gill 1980.

16. Rush et al. 2009; Winker 1994.

17. Toews et al. 2016.

18. Gill 1980. The authors of the 2016 genome study concluded that these species must have a long history of hybridization since so much of the genome has stayed so similar.

19. Leichty and Grier 2006.

20. Mayr and Diamond 2001. Based on mitochondrial DNA, many of these forms are now considered species, and they have been moved from genus *Monarcha* to a new genus, *Symposia-chrus* (Filardi and Smith 2005).

21. Helbig et al. 2002. Tobias et al. (2010a) showed how this might be done in practice for both European and tropical American birds.

22. Cracraft 1989; Zink 1996.

23. Zink 1994.

24. Gill 2014.

25. Two important books on bird speciation are those by Trevor Price (2008, fairly technical) and by Peter R. Grant and B. Rosemary Grant (2008, less technical).

26. This is the approach taken in the most comprehensive recent book on speciation, by Coyne and Orr (2004)

27. Friesen et al. 2007; Taylor et al. 2019.

28. Sorenson et al. 2003.

29. Coyne and Price 2000.

30. Randi and Bernard-Laurent 1999.

31. Saino and Villa 1992; Poelstra et al. 2014; Knief et al. 2019.

32. Grant and Grant 2008, pp. 97–99.

33. Buckley 1969; McQuillan et al. 2018.

34. Price 2008, chapter 16; Uy et al. 2018.

35. Nosil 2012.

36. Tobias et al. 2010b.

37. The finch study is reported by Grant and Grant 2008, chapter 7, the woodcreeper study by Derryberry et al. 2012.

38. Benkman et al. 2009.

39. Sætre et al. 1997; Qvarnström et al. 2010; Sætre and Sæther 2010.

40. Seddon et al. 2013.

41. Vortman et al. 2011; Safran et al. 2016; Hund et al. 2020.

42. Whittingham et al. 2015; Dunn et al. 2013.

43. Price and Bouvier 2002.

44. Freeman et al. 2017. The authors suggest that song learning reduces the rate of evolution because individuals vary so much in what they learn that there is less consistent natural selection for the ability to discriminate the species' song from other songs.

45. Weir and Price 2011.

46. Hermansen et al. 2011, Trier et al. 2014, Barrera-Guzmán et al. 2018.

47. Good et al. 2000.

48. Kleindorfer et al. 2014. Rhymer and Simberloff (1996) summarized what was then known about hybridization as a cause of extinction.

49. Rosenblum et al. 2012.

50. Pigot and Tobias 2013. In the Andes in Peru, pairs of closely related species that defend year-round territories do not have overlapping altitudinal distributions, whereas pairs that lack this behavior do tend to overlap (Freeman et al. 2019).

51. Rabosky and Matute 2013; Cooney et al. 2017b; Harvey et al. 2019.

52. Mason et al. 2016.

Chapter 11. A World of Birds

1. Species numbers from del Hoyo and Collar (2016) *HBW and BirdLife International Illustrated Checklist of the Birds of the World*, vol. 2.

2. Billerman et al. 2011.

3. Barhoum and Burns 2002.

4. Mayr 2004.

5. Louchart et al. 2008.

6. Mayr 2017.

7. Field and Hsiang 2018.

8. Moyle 2005; Prum et al. 2015; Mayr 2017.

9. Among these textbooks was one edition of my own (Futuyma 2005).

10. The names Neognathae ("new jaw," from Greek) and Palaeognathae ("old jaw") refer to the anatomy of the upper jaw and palate.

11. Harshman et al. 2008; Philips et al. 2010; Smith et al. 2013; Baker et al. 2014; Mitchell et al. 2014; Yonezawa et al. 2017; Cloutier et al. 2019.

12. Yonezawa et al. 2017.

13. The diversification and biogeography of Passeriformes is the subject of a new book, *The Largest Avian Radiation*, edited by J. Fjeldså et al. (2020).

14. Sibley and Ahlquist 1990.

15. Ericson et al. 2003; Barker et al. 2004.

16. Oliveros et al. 2019.

17. The names can be confusing. *Corvus* is a genus; Corvidae is the including family; Corvides is a name (of unspecified rank) for the larger clade that includes this and other families.

18. Jønsson et al. 2017.

19. Aggerbeck et al. 2014; Jønsson et al. 2017.

20. Sayol et al. 2019.

21. Beckman and Witt 2015.

22. Barker et al. 2013.

23. Zuccon et al. 2012.

24. Confusingly, the Scarlet, Summer, Hepatic, and Western Tanagers, familiar to North American birders, are no longer in the tanager family but in the Cardinalidae! And the tanager family Thraupidae includes many seedeaters and other species that aren't called tanagers. All these English names are much older than the research that has reclassified these birds.

25. Barker et al. 2015.

26. Weir et al. 2009; Barker et al. 2015.

27. Checklist of the Birds of New York State 2019.

28. Collerton 2012.

29. Fjeldså et al. 2012.

30. Rabosky and Hurlbert 2015.

31. Fjeldså et al. 2018.

32. Kennedy et al. 2014.

33. Jetz and Fine 2012, Pulido-Santacruz and Weir 2016.

34. Weir and Schluter 2007; Pulido-Santacruz and Weir 2016; Schluter 2016.

35. Harvey et al. 2017; Smith et al. 2017.

36. Futuyma 2010; Rosenblum et al. 2012; Dynesius and Jansson 2014; Cutter and Gray 2016.

37. Pulido-Santacruz and Weir 2016.

38. Hawkins et al. 2006.

39. Futuyma 1987; Jansson and Dynesius 2002.

40. Saupe et al. 2019.

41. Fine 2015.

Chapter 12. Evolution and Extinction: The Future of Birds

1. Rosenberg et al. 2019; BirdLife International 2018.

2. Kolbert 2014.

3. Birdlife International. 2018. www.birdlife.org/sites/default/files/attachments/BL_ReportENG_V11_spreeds.pdf.

4. Some species may be more threatened than the IUCN estimate (Ramesh et al. 2017).

5. Species may be affected by more than one threat.

6. Warner 1968; Hempel et al. 2017.

7. Loss et al. 2015.

8. Conrad et al. 2006; Hallmann et al. 2017; Sánchez-Bayo and Wyckhuys 2019; van Klink et al. 2020.

9. Forister et al. 2019.

10. Nebel et al. 2010.

11. Runge et al. 2015.

12. Why they don't—why a species instead remains specialized for a specific habitat—is a complex question that biologists are still working to answer (Futuyma and Moreno 1988). Part of the answer is that natural selection is stronger for maintaining adaptation to the habitat used by most of the population and weaker for adapting to a different habitat that is utilized by fewer individuals (Holt 1996).

13. Wiens and Graham 2005.

14. Miller et al. 2013.

15. Tollefson 2019. Unfortunately, there is little sign that reduction of greenhouse gases will be at all close to the necessary level.

16. MacNally et al. 2009; Iknayan and Beissinger 2018.

17. Riddell et al. 2019.

18. Mayor et al. 2017. The phenological synchrony of all kinds of interactions among species has declined in recent decades (Kharouba et al. 2018).

19. Both et al. 1996; Both and Visser 2006.

20. Jiguet et al. 2010.

21. Freeman et al. 2018.

22. Colwell et al. 2008; Sekercioğlu et al. 2012.

23. Blake and Loiselle 2016. These authors report data from 2001 to 2014; Bette Loiselle told me that the decline has continued into 2020.

24. Deutsch et al. 2008; Huey et al. 2012. These authors propose that tropical species are in greater danger from climate warming than are temperate-zone species.

25. Carlson et al. 2014; Bell 2017.

26. Gienapp et al. 2018.

27. Gardner et al. 2009. The most likely plastic basis for body size is nutrition, which would alter growth rates. The authors showed that there has been no change in the growth rate of feathers, which would be affected by change in nutrition.

28. Socolar et al. 2017.

29. Four decades' data on seventy-three species of birds in Finland also showed that many breed earlier now than in the past (Hällfors et al. 2020).

30. Charmantier et al. 2014; Phillimore et al. 2016.

31. Helm et al. 2019.

32. This is one of many examples of genes that have persisted, and maintained similar function, for almost the entire history of animal evolution. Arthropods (including insects) and chordates (including birds) diverged from a common ancestor more than 540 Ma.

33. Bourret and Garant 2015.

34. Mueller et al. (2011) studied the Blackcap, Liedvogel et al. (2009) the Blue Tit, Ralsten et al. (2019) the Blackpoll Warbler.

35. Nogués-Bravo et al. 2018. Many of these studies are based on fossil pollen rather than larger plant parts.

36. Levine 1998. I started birding in the 1950s and have witnessed these changes.

37. Freeman and Freeman 2014.

38. Tingley et al. 2009.

39. Wiens 2016.

40. Zuckerberg et al. 2009.

41. Quintero and Wiens 2013; also Jezkova and Wiens 2016.

42. Akçakaya et al. 2018.

43. Gerard Manley Hopkins, "Inversnaid."

BIBLIOGRAPHY

Abrahamczyk, S., et al. 2014. Escape from extreme specialization: passionflowers, bats and the sword-billed hummingbird. Proceedings of the Royal Society B 281:20140888.

Abzhanov, A., et al. 2004. *Bmp4* and morphological variation of beaks in Darwin's finches. Science 305:1462–65.

Aggerbeck, M., et al. 2014. Resolving deep lineage divergences in core corvoid passerine birds supports a proto-Papuan island origin. Molecular Phylogeny and Evolution 70:272–85.

Akçakaya, R., et al. 2018. Quantifying species recovery and conservation success to develop an IUCN Green List of Species. Conservation Biology 32:1128–38.

Alfaro, M. E., et al. 2018. Explosive diversification of marine fishes at the Cretaceous–Palaeogene boundary. Nature Ecology and Evolution 2:688–96.

AnAge: The Animal Ageing and Longevity Database. www.genomics.senescence.info.

Anderson, D. J. 1990. Evolution of obligate siblicide in boobies. I. A test of the insurance-egg hypothesis. American Naturalist 135:334–50.

Andersson, M. B. 1982. Female choice selects for extreme tail length in a widowbird. Nature 299:818–20.

Aplin, L. M., et al. 2015. Experimentally induced innovations lead to persistent culture via conformity in wild birds. Nature 518:538–41.

Aplin, L. M., B. Sheldon, and J. Morand-Ferron. 2013. Milk-bottles revisited: social learning and individual variation in the blue tit (*Cyanistes caeruleus*). Animal Behaviour 85:1225–32.

Arnqvist, G., and L. Rowe. 2005. *Sexual Conflict*. Princeton University Press, Princeton, NJ.

Atwell, J. W., et al. 2014. Hormonal, behavioral, and life-history traits exhibit correlated shifts in relation to population establishment in a novel environment. American Naturalist 184: E147–60.

Baiäo, P. C., E. A. Schreiter, and P. G. Parker. 2007. The genetic basis of the plumage polymorphism in Red-footed Boobies (*Sula sula*): a *Melanocortin-1-Receptor* (*MC1R*) analysis. Journal of Heredity 98:287–92.

Baker, A. J., et al. 2014. Genomic support for a moa–tinamou clade and adaptive morphological convergence in flightless ratites. Molecular Biology and Evolution 31:1686–96.

Baker, M. C., and A.E.M. Baker. 1990. Reproductive behavior of female buntings: isolating mechanisms in a hybridizing pair of species. Evolution 44:332–38.

Balakrishnan, C. N., and S. V. Edwards. 2009. Nucleotide variation, linkage disequilibrium and founder-facilitated speciation in wild populations of the zebra finch (*Taeniopygia guttata*). Genetics 181:645–66.

Baldwin, M. W., et al. 2014. Evolution of sweet taste perception in hummingbirds by transformation of the ancestral umami receptor. Science 345:929–33.

Barhoum, D. N., and K. J. Burns. 2002. Phylogenetic relationships of the Wrentit based on mitochondrial cytochrome *b* sequences. Condor 104:740–49.

Barker, F. K., et al. 2004. Phylogeny and diversification of the largest avian radiation. Proceedings of the National Academy of Sciences of the USA 101:11040–45.

Barker, F. K., et al. 2013. Going to extremes: contrasting rates of diversification in a recent radiation of New World passerine birds. Systematic Biology 62:298–320.

Barker, F. K., et al. 2015. New insights into New World biogeography: an integrated view from the phylogeny of blackbirds, cardinals, sparrows, tanagers, warblers, and allies. Auk 132:333–48.

Barrera-Guzmán, A. O., et al. 2018. Hybrid speciation leads to novel male secondary sexual ornamentation of an Amazonian bird. Proceedings of the National Academy of Sciences of the USA 115: E218–25.

Barrowclough, G. F., et al. 2016. How many kinds of birds are there and why does it matter? PLoS ONE 11(11): e0166307. https://doi.org/10.1371/journal.pone.0166307.

Beall, C. M. 2014. Adaptation to high altitude: phenotypes and genotypes. Annuual Review of Anthropology 43:251–72.

Beauchamp, G. 1988. The effect of group size on food intake rate in birds. Biological Reviews of the Cambridge Philosophical Society 73:449–72.

Beckman, E. J., and C. C. Witt. 2015. Phylogeny and biogeography of the New World siskins and goldfinches: rapid, recent diversification in the Central Andes. Molecular Phylogeny and Evolution 87:28–45.

Bell, G. 2017. Evolutionary rescue. Annual Review of Ecology, Evolution, and Systematics 48:605–27.

Benkman, C. W. 1993. Adaptation to single resources and the evolution of crossbill (*Loxia*) diversity. Ecological Monographs 63:305–325.

Benkman, C. W. 2003. Divergent selection drives the adaptive radiation of crossbills. Evolution 57:1176–81.

Benkman, C. W., et al. 2009. A new species of the red crossbill (Fringillidae: *Loxia*) from Idaho. Condor 111:169–76.

Bennett, P. M., and I.P.F. Owens. 2002. *Evolutionary Ecology of Birds*. Oxford University Press, Oxford.

Berglund, A., et al. 1996. Armaments and ornaments: an evolutionary explanation of traits of dual utility. Biological Journal of the Linnaean Society 58:385–99.

Berthold, P. 1995. Microevolution of migratory behaviour illustrated by the blackcap *Sylvia atricapilla*: 1993 Witherby Lecture. Bird Study 42(2): 89–100. https://doi.org/10.1080/00063659509477155.

Bezzel, E., and W. Jetz. 1995. Delay of the autumn migratory period in the blackcap (*Sylvia atricapilla*) 1966–1993—a reaction to global warming. Journal für Ornithologie 136:83–87.

Bhullar, B.-A.S., et al. 2015. A molecular mechanism for the origin of a key evolutionary innovation, the bird beak and palate, revealed by an integrative approach to major transitions in vertebrate history. Evolution 69:1665–77.

Billerman, S. M., et al. 2011. Population genetics of a recent transcontinental colonization of South America by breeding barn wwallows (*Hirundo rustica*). Auk 128:1–8.

BirdLife International. 2018. State of the World's Birds: Taking the Pulse of the Planet. Birdlife International, Cambridge. www.birdlife.org/sites/default/files/attachments/BL _ReportENG_V11_spreeds.pdf.

Birkhead, T., J. Wimpenny, and R. Montgomerie. 2014. *Ten Thousand Birds: Ornithology Since Darwin*. Princeton University Press, Princeton, NJ.

Blake, J. G., and B. Loiselle. 2015. Enigmatic declines in bird numbers in lowland forest of eastern Ecuador may be a consequence of climate change. PEERJ 3: e1177.

Boag, P. T. 1983. The heritability of external morphology in Darwin's ground finches (*Geospiza*) on Isla Daphne Major, Galápagos. Evolution 37:877–94.

Bokony, V., et al. 2006. Multiple cues in status signalliung: the role of wingbars in aggressive interactions of male house sparrows. Ethology 112:947–54.

Bonier, F., et al. 2007. Maternal corticosteroids influence primary offspring sex ratio in a free-ranging passerine bird. Behavioral Ecology 18:1045–50.

Bonneaud, C., et al. 2011. Rapid evolution of disease resistance is accompanied by functional changes in gene expression in a wild bird. Proceedings of the National Academy of Sciences of the USA 108:7866–71.

Boonekamp, J. J., et al. 2014. Reproductive effort accelerates actuarial senescence in wild birds: an experimental study. Ecology Letters 17:599–605.

Borgia, G., and G. F. Bell. 2018. Book review of *The Evolution of Beauty: How Darwin's Forgotten Theory of Mate Choice Shapes the Animal World—and Us*, by R. O. Prum. Animal Behaviour 137:187–88.

Bortolotti, G. R., K. L. Wiebe, and W. M. Iko. 1991. Cannibalism of nestling American kestrels by their parents and siblings. Canadian Journal of Zoology 69:1447–53.

Both, C., et al. 1996. Climate change and population declines in a long-distance migratory bird. Nature 4431:81–83.

Both, C., and M. E. Visser. 2006. Adjustment to climate change is constrained by arrival date in a long-term migrant bird. Nature 411:296–98.

Bourret, A., and D. Garant. 2015. Candidate gene-environment interactions and their relationships with timing of breeding in wild bird population. Ecology Evolution 5(17): 3628–41.

Brennan, P.L.R., et al. 2007. Coevolution of male and female genital morphology in waterfowl. PLoS ONE 2(5): e418.

Brewer, D. 2018. *Birds New to Science; 50 Years of Avian Discoveries*. Christopher Helm, London.

Briskie, J. V., and M. Mackintosh. 2004. Hatching failure increases with severity of population bottlenecks in birds. Proceedings of the National Academy of Sciences of the USA 101:558–61.

Brommer, J. E., et al. 2014. Size differentiation in Finnish house sparrows follows Bergmann's rule with evidence of local adaptation. Journal of Evolutionary Biology 27:737–47.

Brooke, M. D., and N. B. Davies. 1988. Egg mimicry by cuckoos *Cuculus canorus* in relation to discrimination by hosts. Nature 335:630–32.

Brouwer, L., et al. 2012. Helpers at the nest improve late-life offspring performance: evidence from a long-term study and a cross-foster experiment. PLoS ONE 7(4): e33167.

Brown, C. R. 1986. Cliff swallow colonies as information centers. Science 234:83–85.

Brown, C. R. 1988. Social foraging in cliff swallows: local enhancement, risk sensitivity, competition and the avoidance of predators. Animal Behaviour 36:780–92.

Brown, C. R., and M. B. Brown. 1991. Selection of high-quality nests by parasitic cliff swallows. Animal Behaviour 41:457–65.

Brown, C. R., and M. B. Brown. 1998. Intense natural selection on body size and wing and tail asymmetry in cliff swallows during severe weather. Evolution 52:1461–75.

Bruce, M. D. 1999. *Family Tytonidae: Barn-owls.* Vol. 5, *Handbook of the Birds of the World,* edited by J. del Hoyo, A. Elliott, and J. Sargatal. Lynx Edicions, Barcelona.

Brumfield, R. T., et al. 2007. Phylogenetic conservatism and antiquity of a tropical specialization: army-ant-following in the typical antbirds (Thamnophilidae). Molecular Phylogeny and Evolution 45:1–13.

Brusatte, S. L., J. K. O'Connor, and E. D. Jarvis. 2015. The origin and diversification of birds. Current Biology 25: R888–98.

Buchanan, K. L., and A. Cockburn. 2013. Fairy-wrens and their relatives (Maluridae) as model organisms in evolutionary ecology: the scientific legacy of Ian Rowley and Eleanor Russell. Emu 113:i–vii. https://doi.org/10.1071/MUv113n3_ED.

Buckley, P.A. 1969. Disruption of species-typical behavior patterns in F_1 hybrid *Agapornis* parrots. Zeitschrift für Tierpsychologie 26(6): 737–43.

Bumpus, H. C. 1899. The elimination of the unfit as illustrated by the introduced sparrow, *Passer domesticus.* Biology Lectures, Marine Biological Laboratories, Woods Hole, 209–26.

Burleigh, J. G., et al. 2015. Building the avian tree of life using a large scale, sparse supermatrix. Molecular Phylogenetics and Evolution 84:53–63.

Burley, N. T., and R. Symanski. 1998. "A taste for the beautiful": latent aesthetic mate preferences for white crests in two species of Australian grassfinches. American Naturalist 152:792–892.

Burns, K. J., et al. 2014. Phylogenetics and diversification of tanagers (Passeriformes: Thraupidae), the largest radiation of Neotropical songbirds. Molecular Phylogeny and Evolution 75:41–77.

Buttemer, W. A. 1992. Differential overnight survival by Bumpus's house sparrows: an alternate interpretation. Condor 94:944–54.

Carling, M. D., and R. T. Brumfield. 2008. Integrating phylogenetic and population genetic analyses of multiple loci to test species divergence hypotheses in Passerina buntings. Genetics 178:363–77.

Carlson, S. M., et al. 2014. Evolutionary rescue in a changing world. Trends in Ecology and Evolution 29:521–30.

Carrascal, L. M., and E. Moreno. 1992. Proximal costs and benefits of heterospecific social foraging in the great tit, *Parus major.* Canadian. Journal of Zoology 70:1947–52.

Caves, E. M., et al. 2015. Hosts of avian brood parasites have evolved egg signatures with elevated information content. Proceedings of the Royal Society B 282:20150598. https://doi.org/10.1098/rspb.2015.0598.

Chakarov, N., et al. 2008. Fitness in common buzzards at the cross-point of opposite melanin-parasite interactions. Functional Ecology 22:1062–69.

Charmantier, A., et al. 2014. Climate change and timing of avian breeding and migration: evolutionary versus plastic changes. Evolutionary Applications 7:15–28.

Checklist of the Birds of New York State. 2019. New York State Ornithological Association. www.nybirds.org.

Chesser, R. T., et al. 2010. Fifty-first supplement to the American Ornithologists' Union *Check-list of North American Birds*. Auk 127:726–44.

Chesser, R. T. et al. 2017. Fifty-eighth supplement to the American Ornithological Society's *Check-list of North American Birds*. Auk 134:751–73.

Claramunt, S., and J. Cracraft. 2015. A new time tree reveals earth history's imprint on the evolution of modern birds. Science Advances 1(11): e1501005.

Clark, C. J., and R. Dudley. 2009. Flight costs of long, sexually selected tails in hummingbirds. Proceedings of the Royal Society B 276:2109–15.

Clarke, A. 2013. Dinosaur energetics: setting the bounds on feasible physiologies and ecologies. American Naturalist 182:283–97.

Clements Checklist. 2019. See https://www.birds.cornell.edu/clementschecklist/overview -august and J. del Hoyo and N. J. Collar. 2014. *HBW and Birdlife International Illustrated Checklist of the Birds of the World*, vol. 1. Lynx Edicions, Barcelona.

Cloutier, A., et al. 2019. Whole-genome analyses resolve the phylogeny of flightless birds (Palaeognathae) in the presence of an empirical anomaly zone. Systematic Biology 68:937–55.

Clutton-Brock, T. 2007. Sexual selection in males and females. Science 318:1882–85.

Cockburn, A. 2006. Prevalence of different modes of parental care in birds. Proceedings of the Royal Society B 273:1375–83.

Cockburn, A., B. J. Hatchwell, and W. D. Koenig. 2017. Sociality in birds. In *Comparative Social Evolution*, edited by D. R Rubenstein and P. Abbot, 320–52. Cambridge University Press, Cambridge.

Coker, C. R., et al. 2002. Intromittent organ morphology and testes size in relation to mating systems in waterfowl. Auk 119:403–13.

Collerton, A. 2012. A Welsh birder abroad: the story behind a record-breaking big year in New York State. Kingbird 63:2–11.

Colwell, R. K., et al. 2008. Global warming, elevational range shifts, and lowland biotic attrition in the wet tropics. Science 322:258–61.

Conrad, M., et al. 2006. Rapid declines of common, widespread British moths provide evidence of an insect biodiversity crisis. Biological Conservation 132:279–91.

Cook, L. M. 2003. The rise and fall of the *Carbonaria* form of the peppered moth. Quarterly Review of Biology 78:399–417.

Cook, L. M., et al. 2012. Selective bird predation on the peppered moth: the last experiment of Michael Majerus. Biology Letters (London) 8:609–12.

Cooke, F., D. T. Parkin, and R. F. Rockwell. 1988. Evidence of former allopatry of the two color phases of lesser snow geese (*Chen caerulescens caerulescens*). Auk 105:467–79.

Cooney, C. R., et al. 2017a. Mega-evolutionary dynamics of the adaptive radiation of birds. Nature 542. https://doi.org/10.1038/nature21074.

Cooney, C. R., et al. 2017b. Sexual selection, speciation and constraints on geographical range overlap in birds. Ecology Letters 20:863–71.

Cooney, C. R., N. Seddon, and J. A. Tobias. 2016. Widespread correlations between climatic niche evolution and species diversification in birds. Journal of Animal Ecology 85:869–78.

Coyne, J. A., and H. A. Orr. 2004. *Speciation*. Sinauer Associates, Sunderland, MA.

Coyne, J. A., and T. D. Price. 2000. Little evidence for sympatric speciation in island birds. Evolution 54:2166–71.

Cracraft, J. 1989. Speciation and its ontology: the empirical consequences of alternative species concepts for understanding patterns and processes of differentiation. In *Speciation and its Consequences*, edited by D. Otte and J. A. Endler, 29–59. Sinauer, Sunderland, MA.

Cresswell, W. 1994a. Flocking is an effective anti-predator strategy in redshanks, *Tringa totanus*. Animal Behaviour 47:433–42.

Cresswell, W. 1994b. The function of alarm calls in redshanks, *Tringa totanus*. Evolution 47:736–38.

Cutter, A. D., and J. C. Gray. 2016. Ephemeral ecological speciation and the latitudinal biodiversity gradient. Evolution 70:2171–85.

Dakin, R., and R. Montgomerie. 2013. Eye for an eyespot: how iridescent plumage ocelli influence peacock mating success. Behavioral Ecology 24:1048–57.

Darwin, C. R. 1839. *Journal of Researches into the Geology and Natural History of the Countries Visited During the Voyage of H. M. S. "Beagle," under the Command of Captain FitzRoy, R. N. from 1832 to 1836*. Henry Colburn, London. (Commonly known as *The Voyage of the Beagle*.)

Darwin, C. 1859. *On the Origin of Species by Means of Natural Selection, or the Preservation of Favoured Races in the Struggle for Life*. John Murray, London.

Darwin, C. 1871. *The Descent of Man, and Selection in Relation to Sex*. John Murray, London. (Reprinted 1981, Princeton University Press, Princeton, NJ.)

Davies, N. B. 2000. *Cuckoos, Cowbirds, and Other Cheats*. T. and A. D. Poyser, London.

Davies, N. B. 2011. Cuckoo adaptations: trickery and tuning. Journal of Zoology 284:1–14.

Dawkins, R. 1976. *The Selfish Gene*. Oxford University Press, Oxford.

de Bakker, M. G., et al. 2013. Digit loss in archosaur evolution and the interplay between selection and constraints. Nature 500:445–48.

del Hoyo, J., ed. 2020. *All the Birds of the World*. Lynx Edicions, Barcelona.

del Hoyo, J., and N. J. Collar. 2014. *HBW and Birdlife International Illustrated Checklist of the Birds of the World*, vol. 1. Lynx Edicions, Barcelona.

del Hoyo, J., and N. J. Collar. 2016. *HBW and BirdLife International Illustrated Checklist of the Birds of the World.*, vol. 2. Lynx Edicions, Barcelona.

del Hoyo, J., A. Elliott, and J. Sargatal, eds. 1992. *Handbook of Birds of the World*. Vol. 1, *Ostrich to Ducks*. Lynx Edicions, Barcelona.

Dennett, D. C. 1995. *Darwin's Dangerous Idea: Evolution and the Meanings of Life*. Simon and Schuster, New York.

de Queiroz, K. 2007. Species concepts and species delimitation. Systematic Biology 56:879–86.

Derryberry, E. P., et al. 2012. Correlated evolution of beak morphology and song in the Neotropical woodcreeper radiation. Evolution 66:2784–97.

Deutsch, C. A., et al. 2008. Impacts of climate warming on terrestrial ectotherms across latitude. Proceedings of the National Academy of Sciences of the USA 105:6668–72.

Dickinson, E. C., et al. 2014. The Howard and Moore Complete Checklist of the Birds of the World, version 4.0. Accessed from www.howardandmoore.org.

Dijkstra, C., et al. 1990. Brood size manipulations in the kestrel (*Falco tinnunculus*): effects on offspring and parent survival. Journal of Animal Ecology 59:269–85.

Dobzhansky, Th. 1973. Nothing in biology makes sense except in the light of evolution. American Biology Teacher 35:125–29.

Domyan, E. T., and M. D. Shapiro. 2017. Pigeonetics takes flight: evolution, development, and genetics of intraspecific variation. Developmental Biology 427:241–50.

Downs, C. T., et al. 2002. Wax digestion by the lesser honeyguide *Indicator minor*. Comparative Biochemistry and Physiology A 133:124–34.

Ducatez, S., et al. 2020. Behavioural plasticity is associated with reduced extinction risk in birds. Nature Ecology and Evolution. https://doi.org/10.1038/s41559-020-1168-8.

Ducrest, A.-L., L. Keller, and A. Roulin. 2008. Pleiotropy in the melanocortin system, coloration and behavioural syndromes. Trends in Ecology and Evolution 23:502–10.

Dunn, P. O., et al. 2013. MHC variation is related to a sexually selected ornament, survival, and parasite resistance in common yellowthroats. Evolution 67:679–87.

DuVal, E. H. 2007. Adaptive advantages of cooperative courtship for subordinate male lance-tailed manakins. American Naturalist 169:423–32.

Dynesius, M., and R. Jansson. 2014. Persistence of within-species lineages: a neglected control of speciation rates. Evolution 68:923–34.

Eckstrom, J.M.M., et al. 2007. Unusual sex roles in a highly promiscuous parrot: the Greater Vasa Parrot *Caracopsis vasa*. Ibis 149:313–20.

Ellegren, H. 2013. The evolutionary genomics of birds. Annual Review of Ecology, Evolution, and Systematics 44:239–59.

Emaresi, G., et al. 2013. Pleiotropy in the melanocortin system: expression levels of this system are associated with melanogenesis and pigmentation in the tawny owl (*Strix aluco*). Molecular Ecology 22:4915–30.

Emaresi, G., et al. 2014. Melanin-specific life histories. American Naturalist 183:269–80.

Emlen, D. J. 2014. *Animal Weapon: The Evolution of Battle*. Henry Holt, New York.

Emlen, S. T. 1990. White-fronted bee-eaters: helping in a colonially nesting species. In *Cooperative Breeding in Birds: Long-term Studies of Ecology and Behavior,* edited by P. B. Stacey and W. D. Koenig, 487–526. Cambridge University Press, Cambridge.

Emlen, S. T., and P. H. Wrege. 1991. Breeding biology of white-fronted bee-eaters at Nakuru—the influence of helpers on breeder fitness. Journal of Animal Ecology 60:309–26.

Ericson, P.G.P., et al. 2003. Evolution, biogeography, and patterns of diversification in passerine birds. Journal of Avian Biology 34:3–15.

Ericson, P.G.P., et al. 2006. Diversification of Neoaves: integration of molecular sequence data and fossils. Biology Letters 2:543–47.

Everhart, M. 2011. Rediscovery of the *Hesperornis regalis* March 1871 holotype locality indicates an earlier stratigraphic occurrence. Transactions of the Kansas Academy of Science 114:59–69.

Fay, R., et al. 2016. Variation in the age at first reproduction: different strategies or individual quality? Ecology 97:1842–51.

Feduccia, A. 2020. *Romancing Birds and Dinosaurs: Forays in Postmodern Paleontology*. Brown-Walker Press, Irvine, CA.

Feeney, W. E., et al. 2013. Brood parasitism and the evolution of cooperative breeding in birds. Science 342:1506–08.

Feeney, W. E., J. A. Welbergen, and N. E. Langmore. 2014. Advances in the study of coevolution between avian brood parasites and their hosts. Annual Review of Ecology, Evolution, and Systematics 45:227–46.

Field, D. J., and A. Y. Hsiang. 2018. A North American stem turaco, and the complex biogeographic history of modern birds. BMC Biology18:102. https://doi.org/10.1186/s12862-018-1212-3.

Field, D. J., et al. 2018a. Early evolution of modern birds structured by global forest collapse at the end-Cretaceous mass extinction. Current Biology 28:1825–31.

Field, D. J., et al. 2018b. Complete *Ichthyornis* skull illuminates mosaic assembly of the avian head. Nature 557:96–100.

Field, D. J., et al. 2020. Late Cretaceous neornithine from Europe illuminates the origins of crown birds. Nature 579:397–401.

Filardi, C. E., and C. E. Smith. 2005. Molecular phylogenetics of monarch flycatchers (genus *Monarcha*) with emphasis on Solomon Islands endemics. Molecular Phylogeny and Evolution 27:776–88.

Fine, P.V.A. 2015. Ecological and evolutionary drivers of geographic variation in species diversity. Annual Review of Ecology, Evolution, and Systematics 46:369–92.

Fisher, J., and R. A. Hinde. 1951. Further observations on the opening of milk bottles by birds. British Birds 44:393–96.

Fjeldså, J. 1983. Ecological character displacement and character release in grebes Podicipedidae. Ibis 125:463–81.

Fjeldså, J., et al. 2012. The role of mountain ranges in the diversification of birds. Annual Review of Ecology, Evolution, and Systematics 43:249–65.

Fjeldså, J., et al. 2018. Rapid expansion and diversification into new niche space by fluvicoline flycatchers. Journal of Avian Biology 49(3): UNSP e01661.

Fjeldså, J., L. Christidis, and P. Ericson, eds. 2020. *The Largest Avian Radiation: The Evolution of Perching Birds, or the Order Passeriformes.* Lynx Edicions, Barcelona.

Fleischer, R. C., C. E. McIntosh, and C. L. Tarr. 1998. Evolution on a volcanic conveyor belt: using phylogeographic reconstructions and K-Ar-based ages of the Hawaiian Islands to estimate molecular evolutionary rates. Molecular Ecology 7:533–45.

Flower, T. P., M. Gribble, and A. R. Ridley. 2014. Deception by flexible alarm mimicry in an African bird. Science 344:513–16.

Forister, M. L., E. M. Pelton, and S. H. Black. 2019. Declines in insect abundance and diversity: we know enough to act now. Conservation Science and Practice 1: E80.

Forrest, S. C., and R. Naveen. 2000. Prevalence of leucism in pygoscelid penguins of the Antarctic Peninsula. Waterbirds 23:283–85.

Forstmeier, W., et al. 2011. Female extrapair mating behavior can evolve via indirect selection on males. Proceedings of the National Academy of Sciences of the USA 108:10608–13.

Freeman, B. G., et al. 2018. Climate change causes upslope shifts and mountaintop extirpations in a tropical bird community. Proceedings of the National Academy of Sciences of the USA 115:11982–87.

Freeman, B. G., and A.M.C. Freeman. 2014. Rapid upslope shifts in New Guinean birds illustrate strong distributional responses of tropical montane species to global warming. Proceedings of the National Academy of Sciences of the USA 111:4490–94.

Freeman, B. G., G. A. Montgomery, and D. Schluter. 2017. Evolution and plasticity: divergence of song discrimination is faster in birds with innate song than in song learners in Neotropical passerine birds. Evolution 71:2230–42.

Freeman, B. G., J. A. Tobias, and D. Schluter. 2019. Behavior influences range limits and patterns of coexistence across an elevational gradient in tropical birds. Ecography 42:1832–40.

Friedman, N. R., et al. 2009. Correlated evolution of migration and sexual dichromatism in the New World orioles (*Icterus*). Evolution 63:3269–74.

Friesen, V. L., et al. 2007. Sympatric speciation by allochrony in a seabird. Proceedings of the National Academy of Sciences of the USA 104:18589–94.

Futuyma, D. J. 1987. On the role of species in anagenesis. American Naturalist 130:465–73.

Futuyma, D. J. 2005. *Evolution*. Sinauer Associates, Sunderland, MA.

Futuyma, D. J. 2009. *Evolution*, 2nd ed. Sinauer Associates, Sunderland, MA.

Futuyma, D. J. 2010. Evolutionary constraints and ecological consequences. Evolution 64:1865–84.

Futuyma, D. J. 2013. *Evolution*, 3rd ed. Sinauer Associates, Sunderland, MA.

Futuyma, D. J., and M. Kirkpatrick. 2017. *Evolution*, 4th ed. Sinauer Associates, Sunderland, MA.

Futuyma, D. J., and G. Moreno. 1988. The evolution of ecological specialization. Annual Review of Ecology and Systematics 19:207–23.

Galeotti, P., et al. 2003. Colour polymorphism in birds: causes and functions. Journal of Evolutionary Biology 16:635–46.

Galla, S. J., and J. A. Johnson. 2014. Differential introgression and effective size of marker type influence phylogenetic inference of a recently diverged avian group (Phasianidae: *Tympanuchus*). Molecular Phylogeny and Evolution 84:1–13.

Gardner, J. L., et al. 2009. Shifting latitudinal clines in avian body size correlate with global warming in Australian passerines. Proceedings of the Royal Society B 276:3845–52.

Gibbs, H. L., et al. 2000. Genetic evidence for female host-specific races of the common cuckoo. Nature 407:183–86.

Gibson, R., and A. Baker. 2012. Multiple gene sequences resolve phylogenetic relationships in the shorebird suborder Scolopaci (Aves: Charadriiformes). Molecular Phylogeny and Evolution 64:66–72.

Gibson, R. M., J. W. Bradbury, and S. L. Vehrencamp. 1991. Mate choice in lekking sage grouse revisited: the roles of vocal display, female site fidelity, and copying. Behavioral Ecology 2:165–80.

Gienapp, P., et al. 2018. Climate change and evolution: disentangling environmental and genetic responses. Molecular Ecology 17:167–78.

Gill, F. B. 1980. Historical aspects of hybridization between blue-winged and golden-winged warblers. Auk 97:1–18.

Gill, F. B. 2007. *Ornithology*. W. H. Freeman, New York.

Gill, F. B. 2014. Species taxonomy of birds: which null hypothesis? Auk 131:150–61.

Gill, F., and D. Donsker, eds. 2019. IOC World Bird List (v9.2). https://doi.org/10.14344/IOC.ML.9.2.

Godoy-Vitorino, F., et al. 2008. Bacterial community in the crop of the hoatzin, a neotropical folivorous flying bird. Applied Environmental Microbiology 74:5905–12.

Gomez, D., and M. Théry. 2007. Simultaneous crypsis and conspicuousness in color patterns: comparative analysis of a neotropical rainforest bird community. American Naturalist 169: S42–61.

Good, T. P., et al. 2000. Bounded hybrid superiority in an avian hybrid zone: effects of mate, diet, and habitat choice. Evolution 54:1774–83.

Goodale, E., et al. 2020. Mixed company: a framework for understanding the composition and organization of mixed-species animal groups. Biological Reviews 95:889–910.

Götmark, F., and M. Ahlström. 1997. Parental preference for red mouths of chicks in a songbird. Proceedings of the Royal Society B 264:959–62.

Gould, S. J. 1989. *Wonderful Life: The Burgess Shale and the Nature of History*. W. W. Norton, New York.

Gowaty, P. A., and W. C. Bridges. 1991. Nestbox availability affects extra-pair fertilization and conspecific nest parasitism in eastern bluebirds, *Sialia sialis*. Animal Behaviour 41:661–75.

Grant, P. R., and B. R. Grant. 2008. *How and Why Species Multiply: The Radiation of Darwin's Finches*. Princeton University Press, Princeton, NJ.

Grant, P. R., and B. R. Grant. 2014. *40 Years of Evolution*. Princeton University Press, Princeton, NJ.

Green, J. P., et al. 2016. Variation in helper effort among cooperatively breeding bird species is consistent with Hamilton's rule. Nature Communications 7:12663.

Gregory, T. R. 2009. Understanding natural selection: essential concepts and common misconceptions. Evolution: Education and Outreach 2:156–75.

Griesser, M. 2013. Do warning calls boost survival of signal recipients? Evidence from a field experiment in a group-living bird species. Frontiers in. Zoology 10: UNSP 49. https://doi.org/10.1186/1742-9994-10-49.

Griffith, S. C., I.P.F. Owens, and K. Thuman. 2002. Extra pair paternity in birds: a review of interspecific variation and adaptive function. Molecular Ecology 11:2195–2212.

Groenewoud, F., et al. 2018. Subordinate females in the cooperatively breeding Seychelles warbler obtain direct benefits by joining unrelated groups. Journal of Animal Ecology 87:1251–63.

Gustafsson, L. 1987. Interspecific competition lowers fitness in collared flycatchers *Ficedula albicollis*: an experimental demonstration. Ecology 68:291–96.

Gustafsson, L., and W. J. Sutherland. 1988. The costs of reproduction in the collared flycatcher *Ficedula albicollis*. Nature 335:813–15.

Hackett, S. J., et al. 2008. A phylogenomic study of birds reveals their evolutionary history. Science 320:1763–68.

Hällfors, M. H., et al. 2020. Shifts in timing and duration of breeding for 73 bird species over four decades. Proceedings of the National Academy of Sciences of the USA 117:18557–65.

Hallmann, C. A., et al. 2017. More than 75 percent decline over 27 years in total flying insect biomass in protected areas. PLoS ONE 12(10): e0185809. https://doi.org/10.1371/journal.pone.0185809.

Hamilton, W. H., and M. Zuk. 1982. Heritable true fitness and bright birds: a role for parasites? Science 213:384–87.

Hare, R. M., and L. W. Simmons. 2019. Sexual selection and its evolutionary consequences in female animals. Biological Reviews 94:929–56.

Harris, R. B., S. M. Birks, and D. Leache. 2014. Incubator birds: biogeographical origins and evolution of underground nesting in megapodes (Galliformes: Megapodidae). Journal of Biogeography 41:2045–56.

Harshman, J., et al. 2008. Phylogenomic evidence for multiple losses of flight in ratite birds. Proceedings of the National Academy of Sciences of the USA 105:13462–67.

Harvey, M. G., et al. 2017. Positive association between population genetic differentiation and speciation rates in New World birds. Proceedings of the National Academy of Sciences of the USA 114:6328–33.

Harvey, M. G., S. Singhal, and D. L. Rabosky. 2019. Beyond reproductive isolation: demographic controls on the speciation process. Annual Review of Ecology, Evolution, and Systematics 50:75–95.

Hasselquist, D., et al. 1996. Correlation between male song repertoire, extra-pair paternity and offspring survival in the great reed warbler. Nature 381:229–32.

Hatchwell, B. J. 2009. The evolution of cooperative breeding in birds: kinship, dispersal and life history. Philosophical Transactions of the Royal Society B 364:3217–27.

Hatchwell, B. J., et al. 2013. Helping in cooperatively breeding long-tailed tits: a test of Hamilton's rule. Philosophical Transactions of the Royal Society B 369:20130565.

Hawkins, B. A., et al. 2006. Post-Eocene climate change, niche conservatism, and the latitudinal diversity gradient of New World birds. Journal of Biogeography 33:770–80.

Hayman, P., J. Marchant, and T. Prater. 1986. Shorebirds: An Identification Guide to the Waders of the World. Houghton Mifflin, Boston.

Heinsohn, R., et al. 2005. Extreme reversed sexual dichromatism without sex role reversal. Science 309:617–19.

Helbig, A. J., et al. 2002. Guidelines for assigning species rank. Ibis 144:518–25.

Helm, B., et al. 2019. Evolutionary response to climate change in migratory pied flycatchers. Current Biology 29:3714–19.

Hempel, G. E., et al. 2017. Invasive parasites and the fate of Darwin's finches in the Galapagos Islands: the case of the vegetarian finch (Platyspiza crassirostris). Wilson Journal of Ornithology 129:345–49.

Hermansen, J. S., et al. 2011. Hybrid speciation in sparrows I: phenotypic intermediacy, genetic admixture and barriers to gene flow. Molecular Ecology 20:3812–22.

Hill, G. E. 1991. Plumage coloration is a sexually selected indicator of male quality. Nature 350:337–39.

Hill, G. E., and K. L. Farmer. 2005. Carotenoid-based plumage coloration predicts resistance to a novel parasite in the house finch. Naturwissenschaften 92:30–34.

Holland, B., and W. R. Rice. 1998. Perspective: chase-away sexual selection: antagonistic seduction versus resistance. Evolution 52:1–7.

Holt, R. D. 1996. Demographic constraints in evolution: towards unifying the evolutionary theories of senescence and niche conservatism. Evolutionary. Ecology 10:1–11.

Horton, B. M., et al. 2014. Estrogen receptor α polymorphism in a species with alternative behavioral phenotypes. Proceedings of the National Academy of Sciences of the USA 111:1443–48.

Hubbard, J. K., et al. 2010. Vertebrate pigmentation: from underlying genes to adaptive function. Trends in Genetics 26:231–39.

Hudon, J., et al. 2013. Diet-induced plumage erythrism in baltimore orioles as a result of the spread of introduced shrubs. Wilson Journal of Ornithology 125:88–96.

Hudon, J., et al. 2017. Diet explains red flight feathers in yellow-shafted flickers in eastern North America. Auk 134:22–33.

Huey, R. B., et al. 2012. Predicting organismal vulnerability to climate warming: roles of behavior, physiology and adaptation. Philosophical Transactions of the Royal Society B 367:1665–79.

Hund, A. K., et al. 2020. Divergent sexual signals reflect costs of local parasites. Evolution 74:2404–18.

Iknayan, K. J., and S. R. Beissinger. 2018. Collapse of a bird community over the past century driven by climate change. Proceedings of the National Academy of Sciences of the USA 115:8597–602.

James, H. F., et al. 2003. *Pseudopodoces humilis*, a misclassified terrestrial tit (Paridae) of the Tibetan Plateau: evolutionary consequences of shifting adaptive zones. Ibis 145:185–202.

Jansson, R., and M. Dynesius. 2002. The fate of clades in a world of recurrent climate change: Milankovitch cycles and evolution. Annual Review of Ecology, Evolution, and Systematics 33:741–77.

Jarvis, E. D., et al. 2014. Whole-genome analyses resolve early branches in the tree of life of modern birds. Science 346:1320–31.

Jennions, M. D., A. P. Møller, and M. Petrie. 2001. Sexually selected traits and adult survival: a meta-analysis. Quarterly Review of Biology 76:3–36.

Jetz, W., and P.V.A. Fine. 2012. Global gradients in vertebrate diversity predicted by historical area-productivity dynamics and contemporary environment. PLoS Biology 10(3): e1001292.

Jetz, W., and D. R. Rubinstein. 2011. Environmental uncertainty and the global biogeography of cooperative breeding in birds. Current Biology 21:72–78.

Jetz, W., C. H. Sercioğlu, and K. Böhring-Gaese. 2008. The worldwide variation in avian clutch size across species and space. PLoS Biology 6(12):2650–57. https://doi.org/10.1371/journal .pbio.0060303.

Jezkova, T., and J. J. Wiens. 2016. Rates of change in climatic niches in plant and animal populations are much slower than projected climate change. Proceedings of the Royal Society B 283:20182104.

Jiguet, F., et al. 2010. Bird population trends are linearly affected by climate change along species thermal ranges. Proceedings of the Royal Society B 277:3601–08.

Johansson, U. S., et al. 2013. A complete multilocus species phylogeny of the tits and chickadees (Aves, Paridae). Molecular Phylogeny and Evolution 69:852–60.

Johnston, R. F., and R. K. Selander. 1964. House sparrows: rapid evolution of races in North America. Science 144:548–50.

Johnston, R. F., and R. K. Selander. 1971. Evolution in the house sparrow. II. Adaptive differentiation in North American populations. Evolution 25:1–28.

Jones, I. L., and F. M. Hunter. 1998. Heterospecific mating preferences for a feather ornament in least auklets. Behavioral Ecology 9:187–92.

Jones, I. L., and F. M. Hunter. 1999. Experimental evidence for mutual inter- and intrasexual selection favoring a crested auklet ornament. Animal Behaviour 57:521–28.

Jønsson, K. A., et al. 2017. Biogeography and biotic assembly of Indo-Pacific corvoid passerine birds. Annual Review of Ecology, Evolution, and Systematics 48:231–53.

Kachroo, A. H., et al. 2015. Systematic humanization of yeast genes reveals conserved functions and genetic modularity. Science 348:921–25.

Kappers, E. F., et al. 2020. Morph-dependent fitness and directional change of morph frequencies over time in a Dutch population of common buzzards *Buteo buteo*. Journal of Evolutionary Biology 33:1306–15.

Kempenaers, B., G. R. Verheyen, and A. A. Dhondt. 1997. Extrapair paternity in the blue tit (*Parus caeruleus*): female choice, male characteristics, and offspring quality. Behavioral Ecology 8:481–92.

Kennedy, J. D., et al. 2014. Into and out of the tropics: the generation of the latitudinal gradient among New World passerine birds. Journal of Biogeography 41:1746–57.

Kharouba, H. M., et al. 2018. Global shifts in the phenological synchrony of species interactions over recent decades. Proceedings of the National Academy of Sciences of the USA 115:5211–16.

Kipp, F. 1942. Studien über den Vogelzug in Zusammenhang mit Flügelbau und Mauserzyklus. Mitteilungen über der Vogelwelt 35:49–80.

Kleindorfer, S., et al. 2014. Species collapse via hybridization in Darwin's tree finches. American Naturalist 183:325–41.

Knief, U., et al. 2012. QTL and quantitative genetic analysis of beak morphology reveals patterns of standing genetic variation in an estrildid finch. Molecular Ecology 21:3704–17.

Knief, U., et al. 2019. Epistatic mutations under divergent selection govern phenotypic variation in the crow hybrid zone. Nature Ecology and Evolution 3(4): 570–76.

Koenig, W. D., and J. Dickinson, eds. 2004. *Ecology and Evolution of Cooperative Breeding in Birds*. Cambridge University Press, Cambridge.

Koenig, W. D., R. L. Mumme, and F. A. Pitelka. 1995. Patterns and consequences of egg destruction among joint-nesting acorn woodpeckers. Animal Behaviour 50:607–21.

Kokko, H., S. C. Griffith, and S. R. Pryke. 2014. The hawk-dove game in a sexually reproducing species explains a colourful polymorphism of an endangered bird. Proceedings of the Royal Society B 281:20141794. https://doi.org/10.1098/rspb.2014.1794.

Kokko, H., and M. D. Jennions. 2008. Parental investment, sexual selection and sex ratios. Journal of Evolutionary Biology 21:919–48.

Kolbert, E. 2014. *The Sixth Extinction: An Unnatural History*. Henry Holt, New York.

Komdeur, J. 2003. Daughters on request: about helpers and egg sexes in the Seychelles warbler. Proceedings of the Royal Society B 270:3–11.

Kornegay, J. R., J. W. Schilling, and A. C. Wilson. 1994. Molecular adaptation in a leaf-eating bird: stomach lysozyme of the hoatzin. Molecular Biology and Evolution 11:921–28.

Kraaijeveld, K., et al. 2007. The evolution of mutual ornamentation. Animal Behaviour 74:657–77.

Krakauer, A. 2005. Kin selection and cooperative courtship in wild turkeys. Nature 434:69–72.

Krama, T., et al. 2012. You mob my owl, I'll mob yours: birds play tit-for-tat game. Scientific Reports 2:800. https://doi.org/10.1038/srep00800.

Krams, I., et al. 2008. Experimental evidence of reciprocal altruism in the pied flycatcher. Behavioral Ecology and Sociobiology 62:599–605.

Krüger, O., et al. 2001. Maladaptive mate choice maintained by heterozygote advantage. Evolution 55:1207–14.

Kruuk, L.E.B., B. C. Sheldon, and J. Merilä. 2002. Severe inbreeding depression in collared flycatchers (*Ficedula albicollis*). Proceedings of the Royal Society B 269:1581–89.

Ksepka, D. T., et al. 2013. Fossil evidence of wing shape in a stem relative of swifts and hummingbirds (Aves, Pan-Apodiformes). Proceedings of the Royal Society B 280:20130580.

Ksepka, D. T., T. A. Stidham, and T. E. Williamson. 2017. Early Paleocene landbird supports rapid phylogenetic and morphological diversification of crown birds after the K–Pg mass extinction. Proceedings of the National Academy of Sciences of the USA 114:8047–52.

Kuijper, B., et al. 2012. A guide to sexual selection theory. Annual Review of Ecology, Evolution, and Systematics 43:287–311.

Küpper, C., et al.2016. A supergene determines highly divergent male reproductive strategies in the ruff. Nature Genetics 48:79–83.

Lack, D. 1947a. *Darwin's Finches*. Cambridge University Press, Cambridge.

Lack, D. 1947b. The significance of clutch size. Ibis 89:302–52.

Lai, Y.-T., et al. 2019. Standing genetic variation as the predominant source for adaptation of a songbird. Proceedings of the National Academy of Sciences of the USA 116:2152–57.

Lande, R. 1980. Sexual dimorphism, sexual selection, and adaptation in polygenic characters. Evolution 34:292–305.

Lefebvre, L., et al. 1957. Feeding innovation and forebrain size in birds. Animal Behaviour 53:549–60.

Leichty, E. R., and J. W. Grier. 2006. Importance of facial pattern to sexual selection in golden-winged warbler (*Vermivora chrysoptera*). Auk 123:962–99.

Lerner, H.R.L., et al. 2011. Multilocus resolution of phylogeny and timescale in the extant adaptive radiation of Hawaiian honeycreepers. Current Biology 21:1838–44.

Levine, E., ed. 1998. *Bull's Birds of New York State*. Comstock Publishing, Cornell University Press, Ithaca.

Liedvogel, M., et al. 2009. Phenotypic correlates of Clock gene variation in a wild blue tit population: evidence for a role in seasonal timing and reproduction. Molecular Ecology 18:2444–56.

Ligon, J. D. 1999. *The Evolution of Avian Breeding Systems*. Oxford University Press, Oxford.

Ligon, J. D., and P. W. Zwartjes. 1995. Ornate plumage of male red junglefowl does not influence mate choice by females. Animal Behaviour 49:117–25.

Liker, A., R. P. Freckleton, and T. Székely. 2013. The evolution of sex roles in birds is related to adult sex ratio. Nature Communications 4:1587. https://doi.org/10.1038/ncomms2600.

Lim, M.C.W., et al. 2019. Parallel molecular evolution in pathways, genes, and sites in high-elevation hummingbirds revealed by comparative transcriptomics. Genome Biology and Evolution 11:1573–85.

Losos, J. 2017. *Improbable Destinies: How Predictable is Evolution?* Penguin, New York.

Loss, S. R., et al. 2015. Direct mortality of birds from anthropogenic causes. Annual Review of Ecology, Evolution, and Systematics 46:99–120.

Louchart, A., and L. Viriot. 2011. From snout to beak: the loss of teeth in birds. Trends in Ecology and Evolution 26:663–73.

Louchart, A. et al. 2008. Hummingbird with modern feathering: an exceptionally well-preserved Oligocene fossil from southern France. Naurwissenschaften 95:171–75.

Lovászi, P., and C. Moskát. 2004. Break-down of arms race between the red-backed shrike (*Lanius collurio*) and common cuckoo (*Cuculus canorus*). Behaviour 141:245–62.

Lovejoy, A. D. 1936. *The Great Chain of Being: A Study of the History of an Idea*. Harvard University Press, Cambridge, MA.

Lovette, I. J., and E. Bermingham. 1999. Explosive speciation in the New World *Dendroica* warblers. Proceedings of the Royal Society B 266:1629–36.

Lovette, I. J., et al. 2010. A comprehensive multilocus phylogeny for the wood-warblers and a revised classification of the Parulidae (Aves). Molecular Phylogeny and Evolution 57:753–70.

Lovette, I. J., and J. W. Fitzpatrick, eds. 2016. *The Cornell Lab of Ornithology Handbook of Bird Biology*, 3rd ed. John Wiley and Sons, Oxford.

Lowther, J. K. 1961. Polymorphism in the white-throated sparrow, *Zonotrichia albicollis* (Gmelin). Canadian Journal of Zoology 39:281–292.

Loyau, A., et al. 2005. Intra- and intersexual selection for multiple traits in the peacock (*Pavo cristatus*). Ethology 111:810–20.

Lyon, B. E. 1993. Conspecific brood parasitism as a flexible female reproductive tactic in American coots. Animal Behaviour 46:911–28.

Lyon, B. E. 2003. Egg recognition and counting reduce costs of avian conspecific brood parasitism. Nature 422:495–99.

Lyon, B. E., and J. M. Eadie. 2008. Conspecific brood parasitism in birds: a life-history perspective. Annual Review of Ecology, Evolution, and Systematics 39:343–63.

Lyon, B. E., J. M. Eadie, and L. D. Hamilton. 1994. Parental choice selects for ornamental plumage in American coot chicks. Nature 371:240–43.

Lyon, B. E., and R. Montgomerie. 2012. Sexual selection is a form of social selection. Philosophical Transactions of the Royal Society B 367:2266–73.

Lyon, B. E., R. D. Montgomerie, and L. D. Hamilton. 1987. Male parental care and monogamy in snow buntings. Behavioral Ecology and Sociobiology 20:377–82.

MacArthur, R. M. 1958. Population ecology of some warblers of northeastern coniferous forests. Ecology 39:599–619.

Mackie, R. I. 2002. Mutualistic fermentative digestion in the gastrointestinal tract: diversity and evolution. Integrative and Comparative Biology 42:319–26.

MacNally, R., et al. 2009. Collapse of an avifauna: climate change appears to exacerbate habitat loss and degradation. Diversity and Distributions 15:720–30.

Maglianesi, M. A., et al. 2014. Morphological traits determine specialization and resource use in plant-hummingbird networks in the neotropics. Ecology 95:3325–34.

Mallarino, R., et al. 2011. Two developmental modules establish 3-D beak shape variation in Darwin's finches. Proceedings of the National Academy of Sciences of the USA 108:4057–62.

Manegold, A., and T. Töpfer. 2012. The systematic position of *Hemicircus* and the stepwise evolution of adaptations for drilling, tapping and climbing up in true woodpeckers (Picinae,

Picidae). Journal of Zoological Systematics and Evolutionary Research 51:72–82. https://doi .org/10.1111/jzs.12000.

Marrot, P., et al. 2018. Current spring warming as a driver of selection on reproductive timing in a wild passerine. Journal of Animal Ecology 87:754–64.

Martin, P. R., et al. 2017. The outcomes of most aggressive interactions among closely related species are asymmetric. PeerJ 5: e2847. https://doi.org/10.7717/peerj.2847.

Martin, P. R., and T. E. Martin. 2001. Ecological and fitness consequences of species coexistence: a removal experiment with wood warblers. Ecology 82:189–206.

Martin, T. E. 2015. Age-related mortality explains life history strategies of tropical and temperate songbirds. Science 349:966.

Martínez, A. E., et al. 2018. Social information cascades influence the formation of mixed species flocks of ant-following birds in the Neotropics. Animal Behaviour 135:25–35.

Mason, N. A., et al. 2016. Song evolution, speciation, and vocal learning in passerine birds. Evolution 71:786–96.

Mayor, S. J., et al. 2017. Increasing phenological asynchrony between spring green-up and arrival of migratory birds. Scientific. Reports. 7:1902. https://doi.org/10.1038/s41598-017 -02045-z.

Mayr, E. 1942. *Systematics and the Origin of Species.* Columbia University Press, New York.

Mayr, E. 1960. The emergence of evolutionary novelties. In *Evolution After Darwin.* Vol. 1, *The Evolution of Life,* edited by S. Tax, 349–80. University of Chicago Press.

Mayr, E., and J. A. Diamond. 2001. *The Birds of Northern Melanesia: Speciation, Ecology, and Biogeography.* Oxford University Press, Oxford.

Mayr, G. 2004. Old World fossil record of modern-type hummingbirds. Science 304:861–64.

Mayr, G. 2017. *Avian Evolution: The Fossil Record of Birds and Its Paleobiological Significance.* Wiley Blackwell, Chichester, UK.

McCracken, K. G., et al. 2009. Parallel evolution in the major haemoglobin genes of eight species of Andean waterfowl. Molecular Ecology 18:3992–4005.

McGuire, J. A., et al. 2014. Molecular phylogenetics and the diversification of hummingbirds. Current Biology 24:910–16.

McQuillan, M. A., et al. 2018. Hybrid chickadees are deficient in learning and memory. Evolution 72:1155–64.

Michel, A. J., et al. 2018. The gut of the finch: uniqueness of the gut microbiome of the Galápagos vampire finch. Microbiome 6:167.

Michl, G., et al. 2002. Experimental analysis of sperm competition mechanisms in a wild bird population. Proceedings of the National Academy of Sciences of the USA 99:5466–70.

Miller, E. T., et al. 2013. Niche conservatism constrains Australian honeyeater assemblages in stressful environments. Ecology Letters 16:1186–94.

Minias, P., et al. 2015. Wing shape and migration in shorebirds: a comparative study. Ibis 157:528–35.

Minot, E. O., and C. M. Perrins. 1986. Interspecific interference competition—nest sites for blue and great tits. Journal of Animal Ecology 55:331–50.

Mitchell, K. J., et al. 2014. Ancient DNA reveals elephant birds and kiwi are sister taxa and clarifies ratite bird evolution. Science 344:898–900.

Mittelbach, G. G., et al. 2007. Evolution and the latitudinal diversity gradient: speciation, extinction and biogeography. Ecology Letters 10:315–31.

Mock, D. W. 2004. *More Than Kin and Less Than Kind: The Evolution of Family Conflict.* Belknap Press of Harvard University Press, Cambridge, MA.

Moehring, A. J., and J. W. Boughman. 2019. Veiled preferences and cryptic female choice could underlie the origin of novel sexual traits. Biology Letters 15:20160878.

Møller, A. P. 1988. Badge size in the house sparrow *Passer domesticus.* Effects of intra- and intersexual selection. Behavioral Ecology and Sociobiology 22:373–78.

Møller, A. P. 2001. Heritability of arrival date in a migratory bird. Proceedings of the Royal Society B 268:203–6.

Moreau, R. E. 1944. Clutch size: a comparative study, with reference to African birds. Ibis 86:286–347.

Moyle, R. G. 2005. Phylogeny and biogeographical history of Trogoniformes, a pantropical bird order. Biological Journal of the Linnaean Society 84:725–38.

Muchhala, N., and J. D. Thomson. 2009. Going to great lengths: selection for long corolla tubes in an extremely specialized bat-flower mutualism. Proceedings of the Royal Society B 276:2147–52.

Mueller, J. C., et al. 2011. Identification of a gene associated with avian migratory behavior. Proceedings of the Royal Society B 278:2848–56.

Mulder, R. A., et al. 1994. Helpers liberate female fairy-wrens from constraints on extra-pair mate choice. Proceedings of the Royal Society B 255:223–29.

Muñoz, A. P., et al. 2018 Age effects on survival of Amazon forest birds and the latitudinal gradient in bird survival. Auk 135:299–313.

Natarajan, C., et al. 2018. Molecular basis of hemoglobin adaptation in the high-flying bar-headed goose. PLoS Genetics 14 (4): e1007331. https://doi.org/10.1371/journal.pgen.1007331.

Navara, K. J. 2013. Hormone-mediated adjustment of sex ratio in vertebrates. Integrative and Comparative Biology 53:877–87.

Nebel, S., et al. 2010. Declines of aerial insectivores in North America follow a geographic gradient. Avian Conservation and Ecology 5:1. https://doi.org/10.5751/ACE-00391-050201.

Nice, M. M. 1937. Studies in the life history of the song sparrow. I. Transactions of the Linnaean Society of New York 4:1–247.

Nogués-Bravo, D., et al. 2018. Cracking the code of biodiversity responses to past climate change. Trends in Ecology and Evolution 33:7654–776.

Nosil, P. 2012. *Ecological Speciation.* Oxford University Press, Oxford.

Oliveros, C. H., et al. 2019. Earth history and the passerine superradiation. Proceedings of the National Academy of Sciences of the USA 116:7916–25.

Olson, V. A., et al. 2009. Are parental trade-offs in shorebirds driven by parental investment or sexual selection? Journal of Evolutionary Biology 22:672–82.

Ostrom, J. H. 1975. The origin of birds. Annual Review of Earth and Planetary Sciences 3:35–57.

Ostrom, J. H. 1976. *Archaeopteryx* and the origin of birds. Biological Journal of the Linnaean Society 8:91–182.

Ottenburghs, J., et al. 2016. A tree of geese: a phylogenomic perspective on the evolutionary history of true geese. Molecular Phylogeny and Evolution 101:303–313. https://doi.org /10.1016/j.ympev.2016.05.021.

Padian, K., and L. M. Chiappe. 1998. The origin and early evolution of birds. Biological Reviews 73:1–42.

Parker, T. H. 2003. Genetic benefits of mate choice separated from differential maternal investment in red junglefowl (*Gallus gallus*). Evolution 57:2157–65.

Parker, T. H., and J. D. Ligon, 2007. Multiple aspects of condition influence a heritable sexual trait: a synthesis of evidence for capture of genetic variance in red junglefowl. Biological Journal of the Linnaean Society 92:651–60.

Parker, W. K. 1862. On the osteology of gallinaceous birds and tinamous. Transactions of the Zoological Society of London 5.

Parkes, K. C. 1951. The genetics of the golden-winged × blue-winged warbler complex. Wilson Bulletin 63:5–15.

Passarotto, A., et al. 2018. Colour polymorphism in owls is linked to light variability. Oecologia 187:61–73.

Patricelli, G. L., et al. 2018. Book review of *The Evolution of Beauty*, by R. O. Prum. Evolution 73:115–124.

Pauw, A., et al. 2008. Flies and flowers in Darwin's race. Evolution 63:268–79.

Payne, R. B., an M. D. Sorensen. 2005. *The Cuckoos*. Oxford University Press, Oxford.

Perrins, C. M. 1964. Survival of young swifts in relation to brood-size. Nature 201:1147–48.

Petrie, M., et al. 1991. Peahens prefer peacocks with elaborate trains. Animal Behaviour 41:323–31.

Petrie, M., et al. 1992. Multiple mating in a lekking bird—why do peahens mate with more than one male and with the same male more than once? Behavioral Ecology and Sociobiology 31:349–58.

Petrie, M., and T. Halliday. 1994. Experimental and natural changes in the peacock's (*Pavo cristatus*) train can affect mating success. Behavioral Ecology and Sociobiology 35:213–17.

Pfennig, D. W., and K. S. Pfennig. 2012. *Evolution's Wedge: Competition and the Origins of Diversity*. University of California Press, Berkeley.

Phillimore, A. B., et al. 2016. Passerines may be sufficiently plastic to track temperature-mediated shifts in optimum lay dates. Global Change Biology 22:3259–72.

Phillips, M. J., et al. 2010. Tinamous and moas flock together: mitochondrial genome sequence analysis reveal independent losses of flight among ratites. Systematic Biology 59:90–107.

Pierce, R. J. 1979. Foods and feeding of the wrybill (*Anarhynchus frontalis*) on its riverbed breeding grounds. Notornis 26:1–21.

Pigot, A. L., and J. A. Tobias. 2013. Species interactions constrain geographic expansion over evolutionary time. Ecology Letters 16:330–38.

Place, A. R. 1992. Comparative aspects of lipid digestion and absorption—physiological correlates of wax ester digestion. American Journal of Physiology 263: R464–71.

Place, A. R., and E. W. Stiles. 1992. Living off the wax of the land—bayberries and yellow-rumped warblers. Auk 109:334–45.

Plummer, K. E., et al. 2015. Is supplementary feeding in gardens a driver of evolutionary change in a migratory bird species? Global Change Biology 21:4353–63. https://doi.org/10.1111 /gcb.13070.

Poelstra, J. W., et al. 2014. The genomic landscape underlying phenotypic integrity in the face of gene flow in crows. Science 344:1410–14.

Potti, J., and D. Canal. 2011. Heritability and genetic correlation between the sexes in a songbird sexual ornament. Heredity 106:945–54.

Price, T. 2008. *Speciation in Birds*. Roberts and Company, Greenwood Village, CO.

Price, T. D., and M. M. Bouvier. 2002. The evolution of F_1 postzygotic incompatibilities in birds. Evolution 56:2083–89.

Prince, P. A., et al. 1992. Satellite tracking of wandering albatrosses (*Diomedea exulans*) in the South Atlantic. Antarctic Science 4:31–35.

Prokop, Z., et al. 2012. Meta-analysis suggests choosy females get sexy sons more than "good genes." Evolution 66:2665–73.

Prum, R. O. 1999. Development and evolutionary origin of feathers. Journal of Experimental Zoology 285:291–306.

Prum, R. O. 2017. *The Evolution of Beauty: How Darwin's Forgotten Theory of Mate Choice Shapes the Animal World—and Us*. Doubleday, New York.

Prum, R. O., and A. H. Brush. 2002. The evolutionary origin and diversification of feathers. Quarterly Review of Biology 77:261–95.

Prum, R. O., et al. 2015 A comprehensive phylogeny of birds (Aves) using targeted next-generation sequencing. Nature 526:569–73.

Pryke, S. R., and S. C. Griffith. 2009. Socially mediated trade-offs between aggression and parental effort in competing color morphs. American Naturalist 174:455–64.

Pulido, F., and P. Berthold. 2010. Current selection for lower migratory activity will drive the evolution of residency in a migratory bird population. Proceedings of the National Academy of Sciences of the USA 107:7341–46.

Pulido, F., et al. 2001. Heritability of the timing of autumn migration in a natural bird population. Proceedings of the Royal Society B 268:953–59.

Pulido-Santacruz, P., and J. T. Weir. 2016. Extinction as a driver of avian latitudinal diversity gradients. Evolution 70:860–72.

Quinn, J. L., and W. Cresswell. 2006. Testing domains of danger in the selfish herd: sparrowhawks target widely spaced redshanks in flocks. Proceedings of the Royal Society B 273:2521–26.

Quinn, T. W., G. F. Shields, and A. C. Wilson. 1991. Affinities of the Hawaiian goose based on two types of mitochondrial DNA data. Auk 108:585–93.

Quintero, I., and J. J. Wiens. 2013. Rates of projected climate change dramatically exceed past rates of climatic niche evolution among vertebrate species. Ecology Letters 16:1095–1103.

Qvarnström, A., A. M. Rice, and H. Ellegren. 2010. Speciation in *Ficedula* flycatchers. Philosophical Transactions of the Royal Society B 365:1841–52.

Rabosky, D. L., and A. H. Hurlbert. 2015. Species richness at continental scales is dominated by ecological limits. American Naturalist 185:572–83.

Rabosky, D. L., and D. R. Matute. 2013. Macroevolutionary speciation rates are decoupled from the evolution of intrinsic reproductive isolation in *Drosophila* and birds. Proceedings of the National Academy of Sciences of the USA 110:15354–59.

Ralsten, J., et al. 2019. Length polymorphisms at two candidate genes explain variation of migratory behaviors in blackpoll warblers. Ecology and Evolution 9:8840–45.

Ramesh, V., et al. 2017. IUCN greatly underestimates threat levels of endemic birds in the Western Ghats. Biological Conservation 210:205–21.

Randi, E., and A. Bernard-Laurent. 1999. Population genetics of a hybrid zone between the red-legged partridge and rock partridge. Auk 116:324–37.

Ratcliffe, L. M., and P. R. Grant. 1983. Species recognition in Darwin's finches (*Geospiza, Gould*). I. Discrimination by morphological cues. Animal Behaviour 31:1139–53.

Reed, T. E., et al. 2008. Reproductive senescence in a long-lived seabird: rates of decline in late-life performance are associated with varying costs of early reproduction. American Naturalist 171: E89–101.

Remeš, V., et al. 2015. The evolution of parental cooperation in birds. Proceedings of the National Academy of Sciences of the USA 112:13603–13608.

Remeš, V., B. Matysioková, and A. Cockburn. 2012. Long-term and large-scale analyses of nest predation patterns in Australian songbirds and a global comparison of nest predation rates. Journal of Avian Biology 43:435–44.

Rensch, B. 1947. *Neuere Probleme der Abstammungslehre*. Ferdinand Enke Verlag, Stuttgart.

Rensch, B. 1959. *Evolution above the Species Level*. Columbia University Press, New York.

Reyer, H.-U. 1990. Pied Kingfishers: ecological causes and reproductive consequences of cooperative breeding. In *Cooperative Breeding in Birds*, edited by P. B. Stacey and W. D. Koenig, 527–57. Cambridge University Press, Cambridge.

Rhymer, J. M., and D. Simberloff. 1996. Extinction by hybridization and introgression. Annual Review of Ecology and Systematics 27:83–100.

Richerson, P. J., and R. Boyd. 2005. *Not by Genes Alone: How Culture Transformed Human Evolution*. University of Chicago Press, Chicago.

Riddell, E. A., et al. 2019. Cooling requirements fueled the collapse of a desert bird community from climate change. Proceedings of the National Academy of Sciences of the USA. 116:21609–15.

Riehl, C. 2011. Living with strangers: direct benefits favour non-kin cooperation in a communally nesting bird. Proceedings of the Royal Society B 278:1728–35.

Riehl, C. 2013. Evolutionary routes to non-kin cooperative breeding in birds. Proceedings of the Royal Society B 280:20132245. https://doi.org/10.1098/rspb.2013.2245.

Riehl, C., and M. J. Strong. 2018. Stable social relationships between unrelated females increase individual fitness in a cooperative bird. Proceedings of the Royal Society B 285:20160130. https://doi.org/10.1098/rspb.2018.0130.

Riehl, C., and M. J. Strong. 2019. Social parasitism as an alternative reproductive tactic in a cooperatively breeding cuckoo. Nature 567:96–99.

Robillard, A., et al. 2016. Pulsed resources at tundra breeding sites affect winter irruption at temperate latitudes of a top predator, the snowy owl. Oecologia 181:423–33.

Robinson, S. K., and J. Terborgh. 1995. Interspecific aggression and habitat selection by Amazonian birds. Journal of Animal Ecology 64:1–11.

Robinson, W. D., et al. 2010. Diversification of life histories in New World birds. Auk 127:253–62.

Rosenberg, K. V., et al. 2019. Decline of the North American avifauna. Science 366:120–24.

Rosenblum, E. B., et al. 2012. Goldilocks meets Santa Rosalia: an ephemeral speciation model explains patterns of diversification across time scales. Evolutionary Biology 39:255–61.

Rothstein, S. I. 1990. A model system for coevolution: avian brood parasitism. Annual Review of Ecology and Systematics 21:481–508.

Roulin, A. 2004. The evolution, maintenance and adaptive function of genetic colour polymorphism in birds. Biological Reviews 79:815–48.

Roulin, A., and A.-L. Ducrest. 2013. Genetics of colouration in birds. Seminars in Cell and Developmental Biology 24:594–608.

Roulin, A., et al. 2001. Female- and male-specific signals of quality in the barn owl. Journal of Evolutionary Biology 255:266.

Rowe, L. E., and D. Houle. 1996. The lek paradox and the capture of genetic variance by condition dependent traits. Proceedings of the Royal Society B 263:1415–21.

Runge, C. A., et al. 2015. Protected areas and global conservation of migratory birds. Science 350:1255–58.

Rush, A. C., R. J. Cannings, and D. E. Irwin. 2009. Analysis of multilocus DNA reveals hybridization in a contact zone between *Empidonax* flycatchers. Journal of Avian Biology 40:614–24.

Ryan, M. J. 2018. *A Taste for the Beautiful: The Evolution of Attraction.* Princeton University Press, Princeton, NJ.

Sætre, G.-P., et al. 1997. A sexually selected character displacement in flycatchers reinforces reproductive isolation. Nature 387:589–92.

Sætre, G.-P., and S. A. Saether. 2010. Ecology and genetics of speciation in *Ficedula* flycatchers. Molecular Ecology 19:1091–1106.

Safran, R. J., et al. 2016. The maintenance of phenotypic divergence through sexual selection: an experimental study in barn swallows *Hirundo rustica*. Evolution 70:2074–84.

Saino, N., and S. Villa. 1992. Pair composition and reproductive success across a hybrid zone of carrion crows and hooded crows. Auk 109:543–55.

Sánchez-Bayo, F., and K.A.G. Wyckhuys. 2019. Worldwide decline of the entomofauna: a review of its drivers. Biological Conservation 232:8–27.

Saupe, E. E., et al. 2019. Climatic shifts drove major contraction in avian latitudinal distributions throughout the Cenozoic. Proceedings of the National Academy of Sciences of the USA 116:12895–900.

Sayol, F., et al. 2019. Larger brains spur species diversification in birds. Evolution 73:2085–93.

Schluter, D. 2000. *The Ecology of Adaptive Radiation.* Oxford University Press, Oxford.

Schluter, D. 2015. Speciation, ecological opportunity, and latitude. American Naturalist 187:1–18.

Schondube, J. E., and C. Martínez del Rio. 2003. The flowerpiercers' hook: an experimental test of an evolutionary trade-off. Proceedings of the Royal Society B 270:195–98.

Schroeder, J., et al. 2009. A possible case of contemporary selection leading to decrease in sexual plumage dimorphism in a grassland-breeding shorebird. Behavioral Ecology 20:797–807.

Seddon, N., et al. 2013. Sexual selection accelerates signal evolution during speciation in birds. Proceedings of the Royal Society B 280:20131065.

Seebacher, F. 2003. Dinosaur body temperatures: the occurrence of endothermy and ectothermy. Paleobiology 29:105–22.

Sekercioğlu, C. H., et al. 2012. The effects of climate change on tropical birds. Biological Conservation 148:1–18.

Shakya, S. B., et al. 2017. Tapping the woodpecker tree for evolutionary insight. Molecular Phylogeny and Evolution 116:182–91.

Sheldon, B. C., and H. Ellegren. 1999. Sexual selection resulting from extrapair paternity in collared flycatchers. Animal Behaviour 57:285–98.

Shultz, A. J., and K. J. Burns. 2017. The role of sexual and natural selection in shaping patterns of sexual dichromatism in the largest family of songbirds (Aves: Thraupidae). Evolution 71:1061–74.

Sibley, C. G., and J. E. Ahlquist. 1990. *Phylogeny and Classification of Birds: A Study in Molecular Evolution*. Yale University Press, New Haven, CT.

Silva, C.N.S., et al. 2017. Insights into the genetic architecture of morphological traits in two passerine bird species. Heredity 119:197–205.

Simpson, G. G. 1964. The nonprevalence of humanoids. Science 143:769–75.

Simpson, R. K., et al. 2015. Migration and the evolution of sexual dichromatism: evolutionary loss of female coloration with migration among wood warblers. Proceedings of the Royal Society B 282:20150375.

Skutch, A. F. 1949. Do tropical birds rear as many young as they can nourish? Ibis 91:430–55.

Slagsvold, T., et al. 1988. On the cost of searching for a mate in female pied flycatchers *Ficedula hypoleuca*. Animal Behaviour 36:433–42.

Smeds, L., A. Qvarmström, and H. Ellegren. 2016. Direct estimate of the rate of germline mutation in a bird. Genome Research 26:1211–18.

Smith, B. T., et al. 2017. A latitudinal phylogeographic diversity gradient in birds. PLoS Biology 15 (4): e2001073.

Smith, J.N.M., and A. A. Dhondt. 1980. Experimental confirmation of heritable morphological variation in a natural population of song sparrows. Evolution 34:1155–58.

Smith, J. V., et al. 2013. Ratite nonmonophyly: independent evidence from 40 novel loci. Systematic Biology 62:35–49.

Smith, N. A., and J. A. Clarke. 2015. Systematics and evolution of the Pan-Alcidae (Aves, Charadriiformes). Journal of Avian Biology 45:125–40.

Smith, N. D. 2012. Body mass and foraging ecology predict evolutionary patterns of skeletal pneumaticity in the diverse "waterbird" clade. Evolution 66:1059–78.

Smith, N. D., and D. T. Ksepka. 2015. Five well-supported fossil calibrations within the "waterbird" assemblage (Tetrapoda, Aves). Palaeontologia Electronica 18(1): 7FC.

Smith, T. B. 1990. Natural selection on bill characters in the two bill morphs of the African finch *Pyrenestes ostrinus*. Evolution 44:832–42.

Snowberg, L. K., and C. W. Benkman. 2007. The role of marker traits in the assortative mating within red crossbills, *Loxia curvirostra* complex. Journal of Evolutionary Biology 20:1924–32.

Socolar, J. B., et al. 2017. Phenological shifts conserve thermal niches in North American birds and reshape expectations for climate-driven range shifts. Proceedings of the National Academy of Sciences of the USA 114:12976–981.

Sol, D., et al. 2005. Big brains, enhanced cognition, and response of birds to novel environment. Proceedings of the National Academy of Sciences of the USA 102:5460–65.

Soler, M. 2014. Long-term coevolution between avian brood parasites and their hosts. Biological Reviews 89:688–704.

Sorenson, M. D. 1991. The functional significance of parasitic egg laying and typical nesting in redhead ducks: an analysis of individual behaviour. Animal Behaviour 42:771–796.

Sorenson, M. D., K. M. Sefc, and R. T. Payne. 2003. Speciation by host switch in brood parasitic indigobirds. Nature 424:928–31.

Spottiswoode, C. N., et al. 2011. Ancient host specificity within a single species of brood parasitic bird. Proceedings of the National Academy of Sciences of the USA 108:17738–742.

Spottiswoode, C. N., and M. Stevens. 2010. Visual modeling shows that avian host parents use multiple visual cues in rejecting parasitic eggs. Proceedings of the National Academy of Sciences of the USA 107:8672–76.

Spottiswoode, C. N., and M. Stevens. 2012. Host-parasite arms races and rapid changes in bird egg appearance. American Naturalist 179:633–48.

Spurgin, L. G., et al. 2014. Museum DNA reveals the demographic history of the endangered Seychelles warbler. Evolutionary Applications 7:1134–43.

Sridhar, H., G. Beauchamp, and K. Shanker. 2009. Why do birds participate in mixed-species foraging flocks? A large-scale synthesis. Animal Behaviour 78:337–47.

Stacey, P. B., and W. K. Koenig, eds. 1990. *Cooperative Breeding in Birds: Long-term Studies of Ecology and Behavior*. Cambridge University Press, Cambridge.

Stearns, S. C. 1992. *The Evolution of Life Histories*. Oxford University Press, Oxford.

Stresemann, E. 1919. Über die europäischen Baumläufer. Verhandlungen der Ornithologischen Gesellschaft Bayern 14:39–74. (Cited in Mayr 1942.)

Studd, M. V., and R. J. Robertson. 1985. Evidence for reliable badges of status in territorial yellow warblers (*Dendroica petechia*). Animal Behaviour 33:1102–13.

Sultan, S. E. 2015. *Organism and Environment: Ecological Development, Niche Construction, and Adaptation*. Oxford University Press, Oxford.

Sumasgutner, P., et al. 2016. Family morph matters: factors determining survival and recruitment in a long-lived polymorphic raptor. Journal of Animal Ecology 85:1043–55.

Székely, T. 1996. Brood desertion in Kentish plover *Charadrius alexandrinus*: an experimental test of parental quality and remating opportunities. Ibis 138:749–55.

Székely, T., and I. C. Cuthill. 1999. Brood desertion in Kentish plover: the value of parental care. Behavioral Ecology 10:191–97.

Székely, T., I. C. Cuthill, and J. Kis. 1999. Brood desertion in Kentish plover: sex differences in remating opportunities. Behavioral Ecology 10:185–90.

Székely, T., et al. 2004. Brood sex ratio in the Kentish plover. Behavioral Ecology 15:58–62.

Tarka, M., et al. 2014. Intralocus sexual conflict over wing length in a wild migratory bird. American Naturalist 183:62–73.

Taylor, R. S., et al. 2019. Cryptic species and independent evolution of allochronic populations within a seabird species complex (*Hydrobates* spp.). Molecular Phylogenetics and Evolution 139. https://doi.org/10.1016/j.ympev.2019.106552.

Temeles, E. J., et al. 2009. Effect of flower shape and size on foraging performance and trade-offs in a tropical hummingbird. Ecology 90:1147–61.

Teplitsky, C., et al. 2008. Bergmann's rule and climate change revisited: disentangling environmental and genetic responses in a wild bird population. Proceedings of the National Academy of Sciences of the USA 105:13492–496.

Tinbergen, N. 1953. *The Herring Gull's World*. Collins, London.

Tingley, M. W., et al. 2009. Birds track their Grinnellian niche through a century of climate change. Proceedings of the National Academy of Sciences of the USA 106:19637–43.

Tobias, J. A., et al. 2010a. Quantitative criteria for species delimitation. Ibis 152:724–46.

Tobias, J. A., et al. 2010b. Song divergence by sensory drive in Amazonian birds. Evolution 64:2820–39.

Tobias, J. A., et al. 2011. Year-round resource defence and the evolution of male and female song in suboscine birds: social armaments are mutual ornaments. Journal of Evolutionary Biology 24:2118–38.

Tobias, J. A., et al. 2012. The evolution of females' ornaments and weaponry: social selection, sexual selection, and ecological competition. Philosophical Transactions of the Royal Society B 367:2274–93.

Toews, D.P.L., et al. 2016. Plumage genes and little else distinguish the genomes of hybridizing warblers. Current Biology 26:2313–18.

Toews, D.P.L., N. R. Hofmeister, and S. A. Taylor. 2017. The evolution and genetics of carotenoid processing in animals. Trends in Genetics 33:171–82.

Tollefson, J. 2019. Clock ticking on climate action. Nature 562:172–73.

Török, J., and L. Tóth. 1999. Asymmetric competition between two tit species: a reciprocal removal experiment. Journal of Animal Ecology 68:338–45.

Towers, M. 2017. Evolution of antero-posterior patterning of the limb: insights from the chick. Genesis 56(1): e23047.

Trier, C. N., et al. 2014. Evidence for mito-nuclear and sex-linked reproductive barriers between the hybrid Italian sparrow and its parent species. PLoS Genetics 10(1): e1004075.

Turbott, E. G. 1970. The wrybill: a feeding adaptation. Notornis 17:25–27.

Tuttle, E. M. 2003. Alternative reproductive strategies in the white-throated sparrow: behavioral and genetic evidence. Behavioral Ecology 14:425–32.

Tuttle, E. M., et al. 2016. Divergence and functional degradation of a sex chromosome-like supergene. Current Biology 26:344–350.

Uy, J.A.C., et al. 2018. Behavioral isolation and incipient speciation in birds. Annual Review of Ecology, Evolution, and Systematics 49:1–24.

Vagasi, C. I., et al. 2016. Morphological adaptations to migration in birds. Evolutionary Biology. 43:48–59.

van Klink, R., et al. 2020. Meta-analysis reveals declines in terrestrial but increases in freshwater insect abundances. Science 368:417–20.

van Rhijn, J. G. 1991. The Ruff. T. & A. D. Poyser, London.

Vargas, A. O., and G. P. Wagner. 2009. Frame-shifts of digit identity in bird evolution and cyclopamine-treated wings. Evolution and Development 11:163–69.

Vehrencamp, S. L., and J. S. Quinn. 2004. Joint laying systems. In Ecology and Evolution of Cooperative Breeding in Birds, edited by W. D. Koenig and J. L. Dickinson, 177–96. Cambridge University Press, Cambridge.

Velmala, W., et al. 2015. Natural selection for earlier male arrival to breeding grounds through direct and indirect effects in a migratory songbird. Ecology and Evolution 5:1205–1213.

von Holt, B. M. et al. 2018. Growth factor gene IGF1 is associated with bill size in the black-bellied seedcracker Pyrenestes ostrinus. Nature Communication 9:4855. https://doi.org/10.1038/s41467-018-07374-9.

Vortman, Y., et al. 2011. The sexual signals of the East Mediterranean barn swallow: a different swallow tale. Behavioral Ecology 22:1344–52.

Wagner, G. P. 2015. *Homology, Genes, and Evolutionary Innovation*. Princeton University Press, Princeton, NJ.

Wagner, G. P., and J. Gauthier. 1999. 1, 2, 3 = 2, 3, 4: a solution to the problem of the homology of digits in the avian hand. Proceedings of the National Academy of Sciences of the USA 96:5111–16.

Wang, M., and G. T. Lloyd. 2016. Rates of morphological evolution are heterogeneous in Early Cretaceous birds. Proceedings of the Royal Society B 283. https://doi.org/10.1098/rspb .2016.0214.

Warner, R. E. 1968. The role of introduced diseases in the extinction of the endemic Hawaiian avifauna. Condor 70:101–20.

Weimerskirch, H. 1992. Reproductive effort in long-lived birds: age-specific patterns of condition, reproduction and survival in the wandering albatross. Oikos 64:464–73.

Wiens, J. J. 2016. Climate-related local extinctions are already widespread among plant and animal species. PLoS Biology https://doi.org/10.101371/journal.pbio.2001104.

Wiens, J. J., and C. H. Graham. 2005. Niche conservatism: integrating evolution, ecology, and conservation biology. Annual Review of Ecology, Evolution, and Systematics 36:519–39.

Weir, J. T., and T. D. Price. 2011. Limits to speciation inferred from times to secondary sympatry and ages of hybridizing species along a latitudinal gradient. American Naturalist 177:462–69.

Weir, J. T., and D. Schluter. 2007. The latitudinal gradient in recent speciation and extinction rates of birds and mammals. Science 315:1574–76.

Weir, J. T., et al. 2009. The great American biotic interchange in birds. Proceedings of the National Academy of Sciences of the USA 106:21737–42.

Werner, T. K., and T. W. Sherry. 1987. Behavioral feeding specialization in *Pinaroloxias inornata*, the "Darwin's Finch" of Cocos Island, Costa Rica. Proceedings of the National Academy of Sciences of the USA 84:5506–10.

West-Eberhard, M. J. 1983. Sexual selection, social competition, and speciation. Quarterly Review of Biology 58:155–83.

West-Eberhard, M. J. 2003. *Developmental Plasticity and Evolution*. Oxford University Press, New York.

Whiten, A. 2019. Cultural evolution in animals. Annual Review of Ecology, Evolution, and Systematics 50:27–48.

Whittingham, L. A., and P. O. Dunn. 2016. Experimental evidence that brighter males sire more extra-pair young in tree swallows. Molecular Ecology 25:3706–15.

Whittingham, L. A., et al. 2015. Different ornaments signal male health and MHC variation in two populations of a warbler. Molecular Ecology 24:1584–95.

Wiemann, J., T.-R. Yang, and M. A. Norell. 2018. Dinosaur egg colour had a single evolutionary origin. Nature 565:555–58.

Wiens, J. D., R. G. Anthony, and E. D. Forsman. 2014. Competitive interactions and resource partitioning between northern spotted owls and barred owls in western Oregon. Wildlife Monographs 185:1–50. https://doi.org/10.1002/wmon.1009.

Williams, G. C. 1966. *Adaptation and Natural Selection*. Princeton University Press, Princeton, NJ.

Winker, K. 1994. Divergence in the mitochondrial DNA of *Empidonax traillii* and *E. alnorum*, with notes on hybridization. Auk 111:710–13.

Winker, K. 2010. Is it a species? Ibis 152:679–82.

Witmer, M. C. 1996. Consequences of an alien shrub on the plumage coloration and ecology of cedar waxwings. Auk 113:735–43.

Witmer, M. C., and C. Martínez del Rio. 2001. The membrane-bound intestinal enzymes of waxwings and thrushes: adaptive and functional implications of patterns of enzyme activity. Physiological and Biochemical Zoology 74:584–93.

Witmer, M. C., and P. J. Van Soest. 1998. Contrasting digestive strategies of fruit-eating birds. Functional Ecology 12:728–41.

Wyles, J. S., J. G. Kunkel, and A. C. Wilson. 1983. Birds, behavior, and anatomical evolution. Proceedings of the National Academy of Sciences of the USA 80:4394–97.

Xu, X., et al. 2014. An integrative approach to bird origins. Science 346:1253293. https://doi.org/10.1126/science.1253293.

Yeh, P. J., and T. D. Price. 2004. Adaptive phenotypic plasticity and the successful colonization of a novel environment. American Naturalist 164:531–42.

Yi, X., et al. 2010. Sequencing of 50 human genomes reveals adaptation to high altitude. Science 329:75–78.

Yonezawa, T., et al. 2017. Phylogenomics and morphology of extinct paleognaths reveal the origin and evolution of the ratites. Current Biology 27:68–77.

Young, B. E. 1994. Geographic and seasonal patterns of clutch-size variation in house wrens. Auk 111:545–55.

Zahavi, A. 1974. Mate selection—selection for a handicap. Journal of Theoretical Biology 53:205–14.

Zhu, X., et al. 2018. Divergent and parallel routes of biochemical adaptation in high-altitude passerine birds from the Qinghai-Tibetan Plateau. Proceedings of the National Academy of Sciences of the USA 115:1865–70.

Zink, R. M. 1994. The geography of mitochondrial DNA variation, population structure, hybridization, and species limits in the fox sparrow (*Passerella iliaca*). Evolution 48:96–111.

Zink, R. M. 1996. Species concepts, speciation, and sexual selection. Journal of Avian Biology 27:1–6.

Zuckerberg, B., et al. 2009. Poleward shifts in breeding bird distributions in New York State. Global Change Biology 15:1866–83.

Zuccon, D., et al. 2012. The phylogenetic relationships and generic limits of finches (Fringillidae). Molecular Phylogeny and Evolution 62:581–96.

Zuk, M., et al. 1995. Endocrine-immune interactions, ornaments and mate choice in red jungle fowl. Proceedings of the Royal Society B 260:205–10.

Zuk, M., et al. 2006. Silent night: Adaptive disappearance of a sexual signal in a parasitized population of field crickets. Biology Letters 2:521–24.

INDEX

Page numbers in italics refer to tables and figures.

A NOTE ON THE TYPE

This book has been composed in Arno, an Old-style serif typeface in the classic Venetian tradition, designed by Robert Slimbach at Adobe.